Management Training in High-Tech and R&D

NATO ASI Series

Advanced Science Institutes Series

A Series presenting the results of activities sponsored by the NATO Science Committee, which aims at the dissemination of advanced scientific and technological knowledge, with a view to strengthening links between scientific communities.

The Series is published by an international board of publishers in conjunction with the NATO Scientific Affairs Division

A Life Sciences **B Physics**	Plenum Publishing Corporation London and New York
C Mathematical and Physical Sciences **D Behavioural and Social Sciences** **E Applied Sciences**	Kluwer Academic Publishers Dordrecht, Boston and London
F Computer and Systems Sciences **G Ecological Sciences** **H Cell Biology** **I Global Environmental Change**	Springer-Verlag Berlin, Heidelberg, New York, London, Paris and Tokyo

PARTNERSHIP SUB-SERIES

1. Disarmament Technologies	Kluwer Academic Publishers
2. Environment	Springer-Verlag / Kluwer Academic Publishers
3. High Technology	Kluwer Academic Publishers
4. Science and Technology Policy	Kluwer Academic Publishers
5. Computer Networking	Kluwer Academic Publishers

The Partnership Sub-Series incorporates activities undertaken in collaboration with NATO's Cooperation Partners, the countries of the CIS and Central and Eastern Europe, in Priority Areas of concern to those countries.

NATO-PCO-DATA BASE

The electronic index to the NATO ASI Series provides full bibliographical references (with keywords and/or abstracts) to more than 50000 contributions from international scientists published in all sections of the NATO ASI Series.
Access to the NATO-PCO-DATA BASE is possible in two ways:

– via online FILE 128 (NATO-PCO-DATA BASE) hosted by ESRIN, Via Galileo Galilei, I-00044 Frascati, Italy.

– via CD-ROM "NATO-PCO-DATA BASE" with user-friendly retrieval software in English, French and German (© WTV GmbH and DATAWARE Technologies Inc. 1989).

The CD-ROM can be ordered through any member of the Board of Publishers or through NATO-PCO, Overijse, Belgium.

Series 4: Science and Technology Policy – Vol. 12

Management Training in Hight-Tech and R&D

Concept for Enterprises under Transition

edited by

C.M. Rob Verkoeyen†
President of TNP Group,
Brielle, The Netherlands

Albina I. Nikkonen
St Petersburg High-Tech Management Training Center – Director General,
St Petersburg, Russia

and

Andrei Fursenko
Regional Foundation for Science and Technology Development,
St Petersburg, Russia

Springer-Science+Business Media, B.V.

Proceedings of the NATO Advanced Research Workshop on
Concept of Management Training for Enterprises in High-Tech, Science and R&D
St Petersburg, Russia
July 1–5, 1996

A C.I.P. Catalogue record for this book is available from the Library of Congress.

Printed on acid-free paper

ISBN 978-90-481-4803-5 ISBN 978-94-015-8810-2 (eBook)
DOI 10.1007/978-94-015-8810-2

Originally published by Kluwer Academic Publishers in 1997.
Softcover reprint of the hardcover 1st edition 1997

Table of Contents

Introduction

Innovative activities on the development of which Russia puts its hopes for the preservation and successful reconstruction of the former USSR applied science are based on three key factors: funding, infrastructure and management. These factors are of equal significance, however, under today's economic conditions in Russian two latter ones become of critical importance.

This determines the range of readers to whom the present book is addressed:

– representatives of official institutions both Russian and international in charge of technological and scientific sphere reorganization, high-tech industry conversion, market transformations under transit economies;

– Western experts, lecturers, consultants participating in the implementation of various bilateral and international programs according to the above mentioned directions;

– Russian specialists establishing new high-tech enterprises and organizations and corresponding innovative infrastructure.

This book brought to readers' attention includes edited presentations at NATO Advanced Research Workshop (ARW) "Concept of Management Training for Enterprises in High-Tech, Science and R&D" (1-5 July, 1996, St.Petersburg, Russia). This seminar, like the two previous ones ("Technology Dissemination and Marketing of High-Tech Know-How", 1994, and "Conversion of Military Enterprises – A Practical Approach of Industry and Science", 1995), held in St.Petersburg under the auspices of NATO Scientific Affairs Division and Ministry of Science and Technical Policy of the Russian Federation (now the State Committee on Science and Technology of the Russian Federation), was not just an academic one. Among the participants in ARW there were both people developing methodology and implementing various approaches to high-tech managers training and specialists in practice (administrators, e conomists and financiers, top managers of enterprises and consulting firms). All of them came having in mind not just the exchange of information, but also setting of new business contacts and ensuring of further cooperation.

A great role in creating the business atmosphere during ARW was played by the Co-Director on NATO side, President of TNP (Trans National Projects) Management Group Drs. C.M. Rob Verkoeyen. At the preparatory stage of ARW he demonstrated not only his remarkable personal business qualities, but also the full trust in his Russian partners which was of great help in solving most complicated problems. ARW being over, he got numerous proposals not only on training, but also on establishing cooperation ties between the organizations and firms of different countries.

We were waiting for Rob to come to St.Petersburg to finish the preparing of this book and to discuss further cooperation in a number of joint projects. Instead of that we got tragic news on the untimely death of this young man full of energy. Let this book be dedicated to his memory.

Gratitude:

The publishers would like to express their deep gratitude to all people who helped us in conducting ARW and preparing the present books. ARW would have never been a success if it had not been for the help of our sponsors whom we would like to thank on behalf of all participants in the seminar. The sponsors are:

- NATO Scientific Affairs Division;
- State Committee for Science and Technology of the Russian Federation;
- Committee for Economics and Industrial Policy of St.Petersburg Administration.

We would like also to convey our thanks to all specialists of St.Petersburg Regional Foundation for Scientific and Technological Development and St.Petersburg High-Tech Management Training Center who shared the heavy burden of troubles in arranging ARW.

The advice and assistance of Dr. Alain H.Jubier, Programme Director, Science and Technology Policy, were invaluable for us in conducting all the three seminars in St.Petersburg as well as the constant attention and support of the Deputy Chairman of the State Committee on Science and Technology of the Russian Federation Mr. Zurab A.Yakobashvili and the First Vice-Governor of St.Petersburg Administration Mr. Dmitri V.Sergeyev.

We would like to thank our friend Dr. Hanns Peter Muth whose friendly concern and professional consultations contribute to the improvement of our qualification during several years already.

All technical work in translations editing and bringing the articles to the condition acceptable for publishing was carried out by Margarita Zotova, Natalja Zhukovskaya and Andrey Sharapov which allowed us to provide a high-quality lay-out of the book within a short time.

Finally we would like to express our gratitude to Mrs. Wil Bruins for her non-formal friendly assistance in preparing the book.

Unfortunately, we cannot enumerate the names of all people having taken part in conducting the seminar and preparing the book, however, we remember all of them and thank them for cooperation.

Albina NIKKONEN

Co-Director NATO ARW
General Director
St.Petersburg High-Tech
Management Training Center

Andrei FURSENKO

Organizing Committee Member
General Director
St.Petersburg Regional
Foundation for S&T Development

Preface

I am very happy to have been invited to attend this NATO Workshop. However, my being here to welcome you means that I am taking the place of someone, whom those of you who know him will be sorry to miss at this Workshop.

Alain Jubier, NATO's Programme Director for Science and Technology Policy, is one of the "Founding Fathers" of these annual workshops organized by the St. Petersburg Foundation for Scientific and Technological Development, and he had looked forward to participating in it this year as he had in the two previous workshops. Alas, a last-minute problem has forced him to cancel his participation with (and I quote) "many regrets indeed". I have therefore been asked to welcome you on NATO's behalf.

Alain regrets his absence because, as he writes in his cancellation message, "the topic is highly important for the development of science and technology in Countries in transition", but also because he will not meet many friends with whom he has enjoyed working, and because 3 July, 1996 is an important date for the Russian Federation.

NATO is particularly pleased and proud that the Ministry of Science and Technology Policy of the Russian Federation and the St. Petersburg Municipality are co-sponsoring this workshop, and considers this a most welcome demonstration of East-West co-operation on an important problem.

Alain personally, and on behalf of NATO's Science Committee, sends his good wishes for a most successful meeting. He hopes for an active participation and believes that all participants will benefit from sharing their experiences with problems of transition from state-planned to market-oriented economies.

It is his wish that personal contacts made at the Workshop will develop into future joint activities and lead to constructive co-operation between scientists engineers and managers, thereby contributing to a successful transition and, ultimately, a more peaceful and secure world.

Alain also wishes to express his gratitude to the two Co-directors of the Workshop and to the Organizing Committee for their valuable work.

Permit me to add a personal note. I, too, am full of expectation for an interesting and stimulating Workshop. I have observed from close-by the hard work of the organisers, many of them will never be seen by the participants, and I want to express my thanks and admiration for their dedicated work.

You will hear a lot about priorities and prioritising as a key management responsibility during the next four days. My compliments to the organizers for having scheduled the meeting so as to allow many participants to avoid conflicts of priorities.

I wonder if you have had any cancellations because Russian television does not show tennis from Wimbledon. But had the Workshop started a week or two earlier, afternoon sessions would have fallen victim to the European football championship; three weeks later, and the Olympics would have done the same damage. So, if you want to know more about prioritizing, ask the workshop organizers.

I wish you all an enjoyable and interesting seminar.

REORGANIZATION OF HIGH-TECH INDUSTRIES: FEDERAL AND REGIONAL ASPECTS

Dmitriy SERGEEV
Deputy Governor of Saint- Petersburg

1. Situation in the S&T sphere

At present, the scientific-technical potential of Russia is divided into four parts:

1) the academic science (the Russian Academy of Science);
2) the Ministry of Science and Technical Politics of the Russian Federation;
3) university science (the RF State Committee on the Higher Education);
4) departmental science (ministries and departments).

One should mention that the scientific-technical politics cannot be apart from the general state policy and the strategy of any structure should be consistent with the logic and demand of the current social-economic situation.

At a certain stage, the decline slow-down and the achievement of a relatively low inflation level made us hope that the Russian economy had entered the depression phase with possible revival of separate production spheres.

The results of the first half of 1996 broke these hopes. The Russian economy entered a new stage of the industry decline. The high-tech production slow-down is 2 to 2.5 times faster than the decline in other spheres of production. First of all, the current crisis affects high technologies. At the same time, one observes a disproportion between various industries, regions, the internal and external markets, between accumulation and consumption.

The irreversible disproportion of important economic and social relations has destructive consequences and cannot be an evidence of any stability in economy. On the contrary, the society has not yet passed the lowest point of the decline and the crisis is approaching the most dangerous stage.

The deep economic crisis of the Russian economy highly affected the scientific-technical complex. The main factors describing the current situation in the scientific-technical sphere are as follows:

1) The unfavourable situation in the macro-economy (inflation, non-payments, high credit rates, etc.). The transition processes cannot contribute at the moment to the efficiency of the scientific-technical complex reorganization.

2) The collapse of the economic relations of the former Soviet Union and the further territorial disintegration of Russia.

C.M.R. Verkoeyen et al. (eds.), Management Training in High-Tech and R&D, 3–16.

The collapse of the Soviet Union broke the existing scientific relations. Russia was left with 60 % of scientific organizations, 70 % of R&D's and design institutes which required 70 % of all expenditures on science. This event affected the general development of science and technology, the existing scientific relationships and aggravated social problems. Now we observe the migration of scientists from former republics to Russia.

Moreover, the relations between research and production enterprises grow weaker inside Russia, as well. This hinders any cooperation in science.

3) Reduction of the commercial production. The production decline (about 55 % of the 1989 level) resulted in the lower demand for new research and technologies. Enterprises have come across the problem of survival and, having no money for R&D's, lose any interest in innovations.

4) Reduction of the state financing. The high-tech sphere cannot become attractive to private investors in the nearest time. That is why, it needs the state support and a step-by-step transfer of applied R&D which are state-financed to the repayable financing basis. The expenditure on science in the GNP profile of our country is 6 to 8 times lower than in 24 advanced countries. Moreover, the money on science comes from the most vulnerable part of the budget provided by taxes. This results in the chronic lack of financing and, consequently, in the growth of the federal debt to science.

5) Strict structure of the scientific-technical complex. The reduction of the federal financing as well as the lowering demand of the usual customers for new R&D's as a result of their paying ability decrease are significantly faster than any possible structural changes in the scientific-technical sphere. This leads to the uncontrollable situation and losses. As for the changed expenditure profile, practically all money is spent on power and salaries and no full-bodied research is possible.

6) Great difference in the scientific-technical level between different economic sectors, departments, regions and inside them. Along with few high-tech production enterprises, there are a lot of technically backward enterprises.

Strong disproportions between the technical levels of different economic sectors appear to be an obstacle to any innovation activity.

7) The lack of the market infrastructure permitting the commercialization of scientific and technical R&D's.

The scientific-technical problems are closely related to the dynamics of the high-tech commercial production, that is why they are influenced by the crisis of Russian industrial enterprises.

Earlier, the departmental science was in the structure of ministries and sectors or production unions and large enterprises. The research and a part of applied R&D's were mainly concentrated in large economic structures, whereas design and production at enterprises and unions.

The situation in the sphere of industrial science is characterized today by the following factors:

1) The hard financial situation at industrial enterprises excludes practically any possibility of the innovation activity. Industrial enterprises are more interested in the modernization of their production and introduce new technologies on a very small scale.

2) Departmental science and technology suffer both from the steep reduction of state financing and from the dismantling of departmental ministries. As at present enterprises try their best to survive and their profitability is going down, they are not able to buy new scientific and technical products.

As a result of the lack of any innovation activity at enterprises, science is mainly financed from the federal budget.

On the whole, the nearest perspectives are as follows:

- practically all researches will be financed by the state and, to a great extent, from the regional budgets;
- the number of the financed R&D/design projects will be considerably reduced;
- separate projects of the fundamental science will be financed in case they require not too much money.

3) During the nearest future, it is necessary to support the reformation of the majority of scientific structures to avoid the uncontrollable decay of science.

The high-tech commercial production as a whole is in a critical situation. Highly-qualified workers, technologists, designers who are able to carry out this or that project at a very high level do not have much work. One can add to this the availability of a modern production infrastructure which is characterized by a resource of stability analogous to the main production.

The U.S. experience shows that the reduction of financing scientific and technical projects in industry has a long-term effect on the innovation process, and the achievement of the initial level of financing takes years. For example, in the 70's, the Americans needed 8 years to recover from the consequences of the steep reduction of financing this kind of projects and to reach the 1969 level. As the production is also affected, investments are low or cease at all, the revival and restructuring of the economy may take whole decades.

The various consequences of the crisis in the scientific-technical sphere evidence that the negative processes affected the very sources of reformation and renewal of the economy. In addition, these processes have a tendency to grow faster, wider and deeper. We still cannot see any signs of their slow-down or localization.

One should also mention that the above considerations represent a very schematic idea about the current events. But it is obvious that if such a development of the economic processes persists, it will result in the irreversible deformation of the financial, material, technical, information and other aspects of the scientific-technical sphere.

2. Federal scientific-technical policy: Distinct priorities are required

The current social-economic situation and the visible tendencies of its dynamics define the importance of the choice of right priorities in the restructuring of the economy. These priorities are to be determined not only in the whole economy, but also in the scientific-technical sphere.

Earlier the country had planned economy, which was the basis for the consistent scientific-technical policy. There existed a very strict mechanism which used a system of scientific-production enterprises. The latter were created to provide the country with new scientific products to make it technically independent and to more closely relate science and production. Their activity covered all key stages from R&D, design and prototype making to commercial production.

Any attempts to follow the same scientific-technical policy in the new conditions of liberalization turned out a failure. The economic programs were mainly directed at the liberalization of the commercial and intermediary activities and none contained any effective methods to provide the production growth.

Neither a clear economic concept nor any conceptual approaches were developed to carry out an efficient scientific-technical policy, although the market economy, particularly at the critical point of its development, requires more attention to the scientific-technical sphere.

It is necessary to develop a strategy which would take into account both the general tendencies of the new social-economic relations and the specifics of the "science-production" alloy.

At present, Russia has neither efficient scientific-technical policy, or clear priorities. At the best, such priorities are determined on an unofficial level.

The absence of priorities, first of all, has a negative effect on the search and use of such R&D's which are on the boundary between different scientific fields or economic branches. But these very R&D's bring the best economic and social results and require more attention, since they can serve as an innovation basis for the country to get out of the crisis.

On the whole, planning and the scientific-technical progress are very important problems. It is obvious that without the state support the modern science cannot go ahead. Long research, experimental and applied stages result in the impossibility of using the available technological and production basis due to the inadequacy of their development. Sometimes the time from the fundamental idea to its practical realization is several decades, moreover scientists cannot determine at once all possible applications of their discoveries. The financing of large scientific-research and experimental projects can be provided only by transnational corporations on the condition of the state support. In such corporations, the share of the state financing of scientific research exceeds 50 % and may be even higher for other areas.

The aggravation of the economic situation in the country and the reduction of the state financing of scientific research make the choice of priorities more urgent.

There are different approaches to solve this problem. For example, the Ministry of Science and Technical Politics think that the reorganization of the scientific-technical potential must be based on a scientific doctrine adequate to the new historical conditions.

However, while developing the doctrine, the Ministry thought unreasonable to determine concrete state priorities of the scientific-technical development.

There are some other approaches to the problem of priorities. According to the opinion of the Russian Academy of Science, it is necessary to concentrate all available resources and direct all efforts at the priorities of the scientific-technical development. These priorities are determined by the crisis of the economy and society and are very important because they can provide the competitiveness of the Russian produce on the internal and external markets and create more qualified jobs.

In our opinion, there is no systematic approach towards the reorganization of the scientific-technical potential: no planning, no rational distribution of funds which is necessary in the current conditions, since the financing of the whole spectrum of scientific research in Russia cannot be afforded.

Therefore, one should identify the priorities to support them in the first place and try to create "breakthrough areas". The state must support priority programs and directions and not concrete scientific-technical organizations. The selection should be very strict, as only those projects which are of primary importance for the country and are based on a considerable scientific resource are preferable.

To determine such priorities, one should use a differential approach to the fundamental research and applied science. The latter which is closer to practice should be "incorporated" into the state anti-crisis policy and correspond to the anti-crisis priorities, i.e. provide the national security as well as the stability in the vital spheres.

On the whole, the analysis of the priority problem brings us to the following conclusions:

1. The determination of the priorities in the Russian scientific-technical policy depends on the possibility of financing of this or that project.

2. There are two main approaches to the problem defined by two possible sources of financing: commercial structures and the state.

It is unlikely that commercial structures will finance a big number of scientific projects. The great majority of such structures are not interested in any production activity. More often they try to buy ready western technologies and equipment and do not want to wait for the results of the domestic projects.

It would seem that the state-financed projects are a different thing. However one may doubt the usefulness of the projects suggested for state financing in the present situation. Generalized definitions in combination with the lack of any detailed description of the approach (criteria) towards project selection cannot give even an approximate idea of how and which projects can be chosen for state financing. Sooner or later, it will be necessary to specify the priorities to get out of the current crisis.

3) One can suggest that in the nearest future (a) the priorities will be determined by the state; (b) to solve the concrete tasks of the scientific research and design, this work will become more of a regional character; (c) as production now has simplified, most of current scientific-research and design projects will be stopped as useless, at the same time simplified scientific and design projects will appear, which will be related with the necessity of the solution of simplified, priority tasks by the available methods.

The experience of many countries evidences that the existence of the market is not a sufficient condition for a fast production growth. The state should undertake resolute and consistent measures to create a favourable innovation climate. This requirement is even more important, as the decline of production, the disorder of the financial and credit system, inflation, non-payments have entailed the unprecedented decrease in investment activities and the lack of stimuli for the innovation efforts.

First of all we should think of the way to organize the high-tech production to get out of the crisis. The most important thing is to develop an absolutely new state doctrine of the scientific-technical development of Russia. This doctrine should comply with the social and state interests, the reform goals and long-term perspectives of the economy.

The urgent measures to maintain the scientific-technical potential and to develop the state doctrine of the scientific-technical development should include both strategical and tactical steps. If no goals are defined, if the goals do not correspond to the means of their achievement, these urgent measures will be a failure.

If to use such an approach and not to wait for some “salvation” from the state, the efforts should be concentrated on the creation of the economic basis for self-reproduction.

But for this purpose, it is necessary to stop the spontaneous and chaotic influence of the market relations on the scientific-technical sphere and to protect the scientific-technical potential against breaking up. The scientific-technical sphere needs some assimilation period to get used to the market relations. This period will include, first, a system of urgent, anti-crisis measures for the maintenance of the scientific-technical potential and, second, the creation of compulsory initial economic, legal, political etc. conditions for the most successful incorporation of the scientific-technical sphere into the market economy. The idea of the assimilation period rejects the way of a radical and speedy transformation of the scientific-technical sphere without any clear understanding of what this sphere is to look like. But this period will bring desired results only if the state follows a very consistent and strong policy. It is obvious that during this period no market methods of self-regulation can be used. The government should develop a strict system of state measures to overcome the crisis and create conditions for the increase in the scientific-technical potential.

The scientific-technical policy of the state is of greatest importance for the scientific production reorganization. It is directed at the development and introduction of new scientific and technical achievements into all spheres of the society through state and non-state structures covering the whole scientific-technical cycle from R&D to the production and use of new produce.

The state policy must have the following goals:

- determination of the priorities of the scientific-technical development and distribution of state funds in accordance with the priorities;
- creation of the innovation climate in economy, the consistent development of the demand for innovations;
- creation of conditions (economic, legal, social) for the development of the business activity in the scientific-technical sphere;
- protection of the intellectual property, the scientific-technical potential of the country.

To realize the scientific-technical policy, both direct and indirect methods of regulation can be used. The state order will remain the main form of the direct regulation both in the state sector and in the sphere of free enterprise for a long time. What is common for the both sectors is their goals, the difference lies in the means used to achieve these goals, the methods of regulation. The state order should be profitable, that is why the state should provide efficient financing, direct subsidies (only for scientific-research and design projects or priority produce), privileged credits. The executor should not have any problems in selling the scientific product.

In addition, the scientific-technical activity is to be regulated via taxes, credits, custom tariffs, subsidies, depreciation policy. It is an absolutely new mechanism for the Russian economy.

One of the urgent tasks of the state is the development of the concept of the consistent inter-regional policy to be carried out in the scientific-technical sphere. To avoid any possibility of shifting regional scientific-technical problems on to local budgets, it is necessary to use high potentials contained in unconventional financial sources originating in the scientific-technical cooperation between regions. The international forms of the scientific-technical cooperation are to undergo some changes, too.

The realization of these and other proposals would make the economic reform more directed at renovation, further development and more efficient use of the scientific-technical potential.

3. Regional aspects of the scientific-technical policy

The economy and natural resources of the regions determine the difference in the structure and specialization of production. These factors are key in the crisis conditions.

The disintegration of the former USSR economy is influenced by several tendencies.

First. Production and sales have become different in different regions due to objective reasons. This tendency was not so pronounced before, as it was smoothed to this or that extent by the state policy.

Second. New economic regulation mechanisms appeared on the local level (prices, tariffs, land plot standards etc.). They are rather different in different regions and often do not agree with the federal legislation. The differences in local legislations increase the isolation of territories from each other, since they arise not from the actual economic differences of the regions, but from the absence of the common market "space".

Third. The processes at the federal level evidence that the government is not able to efficiently control the economy of the regions and does not want to take any responsibilities for the problems they come across letting the local authorities to solve them all by themselves (without enough financial means). In the nearest years, the economy will be characterized by the maximum use of relative advantages of the regions with the purpose of stabilization of their social and economic sphere.

The disintegration of the Russian economy aggravates the economic crisis and poses a number of specific tasks for the structural policy.

These tasks are as follows: a new approach to the regional economy; the integration of the scientific-technical programs and programs of regional development; the identification of priority tasks of the restructuring of the economy.

The new approach to the regional economy consists in:

- search for the rational combination of the methods of the centralized control over the innovation activity and the market self-regulation of the innovation activity in the regions;
- co-ordination of departmental and regional interests in the realization of target scientific-technical programs;
- improvement of the organizational forms of innovation projects.

In our opinion, the regional problems of the high-tech production should be solved in the framework of the industrial policy. The anti-crisis industrial policy is to be directed at the resource mobilization of the region, the solution of vital social problems and the maintenance of the minimally acceptable living standard.

For example, in Moscow, in the framework of the regional industrial policy, they adopted the program "From science to production". The main objective of the program is to support high-tech production in mastering new kinds of produce and services necessary for the city.

The experience shows that the most acceptable organizational form of the combination of departmental and regional interests in the realization of large innovation projects is the target program method which is based on the contractual financing. Money in this case is given to an executor for a concrete research, design etc. works.

To carry out large-scale, regional programs consisting of a number of innovation projects, special managerial bodies should be created to take responsibilities for the achievement of the program goals, the rational spending of the money, the timely fulfillment of the work.

Such an organizational mechanism of management provides the consistency and rationality of the use of the human, natural, financial and scientific-technical resources of the region. In addition, this target program approach suggests that the regional aspect of the scientific-technical policy be in the close relationship with the tasks of the territorial economy and the restructuring of the production.

To carry out the scientific research and development on the local level, which means that the scientific or technical product will have a limited number of customers, it is useful to create scientific-research consortiums (on the basis of existing scientific-research institutes) which would be financed both from their own sources and from the state sources, i.e. by state departments responsible for the scientific-technical development of respective industries and regions.

During a certain period of time, small innovation companies will remain one of the organizational forms of the innovation potential development in regions. At present, we are coming to a better understanding of the objectives of the industry in St. Petersburg. The new approach will be based on a wider use of the industrial and scientific-technical potential.

In this connection, it is planned to carry out an active industrial policy. The Committee on Economy and Industrial Policy is responsible for its development and realization. In the framework of this policy, the Committee will co-ordinate all activities both in the scientific-technical sphere of the industry and in small and middle business. These spheres must take into account the needs of the city; the effective mechanisms are the municipal order, tax holidays and delays, the direct participation of the city in innovation projects, etc.

4. Characterization of the reorganization of high-tech production enterprises

The most typical structure of the high-tech production is a scientific-production union including scientific-technical institutes, design bureaus and plants.

In new market conditions, many enterprises come across a number of problems at the design and pre-production stage. One of the main reasons of these problems is that the Soviet ministries used to supply enterprises with new design products which were created in special design bureaus; in addition, enterprises had guarantied customers. That is why the attempts to enter the market are very painful for such enterprises.

There are also difficulties of the technological nature. They are caused by the absence or the bad organization of the production circuit and the immaturity of the diversification processes (simultaneous, expanding integration). These problems are very characteristic of Russia as the chain "scientific research/design – production – introduction" took a long time, as different, poorly coordinated enterprises took part in it.

In addition, huge expenses incurred at the scientific research and design stage are usually covered after the production of small batches of produce. As a result, the unit produce cost goes steeply up. The price growth, in its turn, affects potential sales and threatens the high-tech production existence.

Thus, the maintenance of scientific-research institutes and design bureaus in the structure of big unions turns out undesirable.

The experience shows that there are some common features in the policy of different enterprises in respect of scientific research and design:

1. Research organizations try to carry out both scientific research and design and even to build some pilot production. This happens because the relations between scientific-research institutes and industrial enterprises tend to break (even in the structure of scientific-production organizations). For example, the end product of the State Scientific-Research Institute for Aviation Systems was the scientific-research product for many years. At present, the institute is creating its own design bureaus to make prototypes for the further small-scale production.

On the other hand, the "Baltiiskii" plant has decided to create a new, modern design bureau. As a rule, ship-building enterprises use the design products of large design bureaus, the role of their own bureaus is usually limited to the adjustment of these products to specific conditions of ship-building projects.

So, the break-up of the relations between the scientific research/design and production units leads to the creation of the whole scientific-design-production circuit at one enterprise.

2. The long time the development and design of a new product takes is unacceptable from the point of view of the customer.

Therefore, it is necessary to reduce the pre-production period. First, the produce will get to the market much faster, and second, the production costs will be steeply reduced.

A number of enterprises try to solve this problem via the automatization of the pre-production stage. In doing so, they try to maintain the same level of the complexity of their produce.

On the whole, the pre-production stage reduction depends on the general simplification of the designed produce and not on the automatization of this stage. The temporary groups can be created to develop the design and production documentation and prepare the product for a commercial production in the shortest time.

Of interest is the experience gained at the "Kirovskii" plant. To reduce the pre-production stage, a part of prototypes were made in the process of commercial production, thus skipping the pilot production.

3. The high-tech production enterprises usually incur great overhead expenses reaching 300 to 600 %. If this level of overhead expenses remains the same for

simplified produce, such an enterprise will not survive in the market conditions. If the simplified produce constitutes only a part of the production, then it needs to be produced on a very large scale, then the profit will be high at a low rate. It is unreal in our conditions.

At present the payable demand for high-quality products may be provided only by the external market. That is why, scientific-research institutes are forced to deal with foreign customers. This creates the problem of the scientific potential leak.

For example, the Research Institute named after Krylov, St. Petersburg, possesses a unique experimental base. The defence production share went down during the last 2 years from 90 to 10–15 %. Most of the developed conversion projects (liquid disinfection instruments, special stuffs for precious stone treatment or cognac artificial ageing acceleration etc.) found no demand on the internal market. Almost all customers of the institute at present are foreign.

The analysis of the experience of Russian enterprises allows one to determine the following major ways of production re-orientation towards other kinds of produce:

1. Re-orientation towards new kinds of produce the production of which can be mastered very quickly and which has the features close to those of the traditionally manufactured produce. In this case, enterprises try not to let any negative changes occur, especially in the highly-qualified research and production personnel.

2. Re-orientation towards simplified produce. If such an approach is used, there is no need in long-term and large-scale reconstruction programs, however some negative tendencies are observed in the social sphere, as the qualification of the personnel goes down. This causes the degradation of the scientific-technical potential.

3. Creation of new areas of activities within the same enterprise (trade, business services, leasing etc.). This requires re-orientation of the personnel towards these new activities and timely measures for their training.

Main task in this case is the maintenance of the reasonable level of diversification and flexible ties between the main and other kinds of activities (the latter concerns also the personnel who must be very mobile).

4. Re-orientation due to the change in the proportion of the available kinds of production (when the traditional production is reduced). The considerable problems in this case are related with:

- the differences in requirements to the personnel of different productions;
- the impossibility (in many cases) to increase one production as a compensation for the reduction of another;
- the high probability that the payable demand for the traditional produce will go down.

However, at present, the activities of enterprises are pre-determined by the dynamics of the crisis tendencies and use specific methods. The anti-crisis behaviour of enterprises (including their personnel) often contradicts the steps which could be efficient and useful in the conditions of the economic growth or stability.

Let us give some characteristics of the anti-crisis policy of industrial enterprises, which influence the speed and degree of their re-orientation. These characteristics are absolutely different from what enterprises “got used” to during the long stability period.

1. The structural economy crisis, first of all, is characterized by the destruction of existing economic relations. At the same time, new economic relations are coming to replace the old ones with a big delay. In these conditions, the specialization of an enterprise and the high level of labour division make the viability of this enterprise much lower. That is why, the naturalization of the production, i.e. scientific research, design and production carried out in the structure of the same enterprise, becomes profitable and allows enterprises reduce losses from having too many people and too small an amount of work.

At present, production cannot buy scientific/design products from other organization, moreover it is not able to finance its own scientific research/design works. Even relatively successful enterprises tend to refuse to buy these products and carry out their own, though simplified, works.

2. The decrease in the income of the majority of the population and their increasing differentiation lead to the decrease in the consumer's demand and to the simplification of the consumption structure. The buying ability of more and more people is going down, respectively, the demand for high-quality and expensive goods is also becoming lower.

This tendency is reflected by the decrease in the volume of high-tech and complex products manufactured by enterprises. This aggravates the problem of the decrease in the qualification of the personnel and simplification of their duties. According to questioning results, more than 50 % of scientists confirm that the general level of research works is steadily going down.

The domestic consumer can hardly be in the need of big batches of high-tech products. In our conditions, major customers are foreign. In this connection, joint-ventures are considered to be one of the most efficient forms.

A joint-venture may improve the technical level of products, but at the same time it may create a lot of problems for the scientific-technical personnel, since they use ready technologies and the task of the personnel is reduced to the mastering of these technologies or to the maintenance of ready equipment.

The economic growth with the simultaneous innovation suggests numerous changes and assimilation to them, innovations undermine the existing production structure. The reformation of the Russian economy changed its structure which was traditionally represented by big, state-owned organizations and unions.

In the current reorganization of industrial enterprises one can clearly see the tendency of the decentralization of their divisions which become financially-independent commercial structures with their own management bodies (profit-centers, business-centers etc.).

During the crisis, the tendency towards the replacement of huge enterprises by small, mobile structures is inevitable. However, the scientific-production level cannot stay the same in these conditions.

On the other hand, it is important to create such a financial and industrial structure which could be maximally protected against unfavourable external factors (e.g. inflation) and various risks (political, economic, social etc.). With this very purpose, high-tech enterprises try to unite financial, research, design and production organizations as well as consumers of their produce. In this case the whole production chain would work perfectly.

5. Reorganization forms of the industrial science

The innovation system of the former Soviet Union was characterized by the hierarchic approach and considerable separateness. Research and design works, experimental and commercial production were considered as separate kinds of activities and were isolated from each other.

This non-market research and innovation system will be difficult to change.

In the long run, industrial innovations start with the analysis of the potential demand and end with the production and sales of competitive goods and services. In the Russian Federation, the innovation measures in the scientific-technical sphere shall meet the priority demands of the anti-crisis policy.

The main problems this approach creates in the specific Russian conditions are:

1. While determining the potential demand: information noise, uncertainty of the tendencies of demand changes among the population, enterprises, the state.
2. While preparing a business-plan: a poor possibility of any quantitative description of restrictions on resources and sales; the increasing role of the business-intuition.
3. During the search for an investor: the absence of market mechanisms to co-ordinate the interests of designers, producers and investors; the economic situation is unfavourable for investments.
4. At the production and sales stage: the restrictions on resources and the buying ability of a customer, his demands are not taken into account.

Thus, the main objective of the new organizational forms of the scientific-technical activity is to provide the joint work of all the members of the innovation chain.

The most important forms are as follows:

1. State scientific centers. By the governmental decision, this status is given to enterprises, scientific organizations and educational institutes for a period of 2 years. The centers have a federal significance and the right to obtain state support of a special kind (target financing, tax holidays, tariff privileges). By 1995, 61 Russian organizations had got this status. The financing of such centers is provided by different state structures: economic departments, the Ministry of Science and Technical Politics, ordering ministries, etc.).
2. Technological parks represent the mechanism of integration of science, production, business, financial structures, local and regional authorities.
3. Financial-industrial groups favour investment processes relating the interests of designers, producers and investors in industry. Financial-industrial groups are of greatest importance for the maintenance of the scientific-technical potential at enterprises manufacturing high-tech products selling well on the world market. These groups include a wide range of research organizations.

   The form of financial-industrial groups is supposed to support large-scale researches and designs, the creation and use of scientific-production complexes, the development of cooperation between enterprises.
4. Business incubators have an objective to provide the economic, scientific-production, information and legal support to incubator groups, temporary collectives and small high-tech business companies.

5. Regional scientific-technical funds. The example of such a form is the successful St. Petersburg Regional Foundation for Scientific & Technical Development.

6. In the associations of enterprises, it is useful to create engineering centers which would have an objective of the introduction of advanced technologies into the consumer's goods production. They could gather all production information of the world, train technical personnel. The engineering centers could transfer the results of scientific-technical discoveries and inventions to the production in a form acceptable for the mass and commercial production.

In mid-1995, a group of small and middle scientific-technical enterprises in conjunction with one of the largest electronic enterprise "AO Svetlana" proposed to create the Innovation Technological Centre. This uncommercial organization unites the efforts of enterprises to promote the Russian high-tech produce on the market and tries to maintain and create new jobs for scientists and highly-qualified specialists and workers.

7. In many countries, wide-spread are technopolises intended for the closer relationship between science and production. The core of technopolises is a big university or several higher education institutes, scientific centers located very close to each other. A scientific centre unites a lot of small high-tech companies (most of them are of a venture nature) as well as intermediary companies providing the technology transfer from the fundamental science to industry.

In our conditions, such complexes are possible in various forms. For example, large defence enterprises may create a favourable environment for small enterprises to appear on their basis and to provide the transfer of high-tech decisions to the civilian production.

8. One of the efficient forms of small creative groups is a venture company. Production brigades, temporary creative groups of design bureaus, small enterprises may be financed on a venture basis.

The analysis of the above-enumerated structures in the scientific-technical sphere shows that they have the following common problems:

- the science-production relationship is very weak (unstable);
- the projects are not always properly justified;
- inter-departmental interactions are rather poor.

The research institutes and enterprises of St. Petersburg do not co-ordinate their work on the development of consistent approaches towards such important aspects of innovation as marketing research for scientific-technical products, information supply, business planning. This results in great expenses the institutes and enterprises incur. These expenses could be essentially reduced if the resources of these institutes and enterprises were combined and if they had a common infrastructure which could solve some regional problems related with the commercialization of the available internal potential of St. Petersburg on the basis of information exchange and wide telecommunication opportunities.

The co-ordination of efforts of various organizations in the development of the complex scientific-technical and innovation policy would also facilitate the task of the city authorities of St. Petersburg and help them to work out the city development strategy in which the realization of the scientific-production potential is one of priority directions. To remove these problems, a regional co-ordination center for innovations and industry development is being created in St. Petersburg by the Association of Industrial Enterprises, among others.

The main objectives of this Centre are:

- methodological support of the city authorities in the co-ordination of the scientific-technical and innovation activities in St. Petersburg;
- preparation of proposals for the focusing of scientific-research resources on the most efficient realization of separate priority programs of the scientific-technical development of the city;
- co-ordination of marketing researches in the field of high technologies;
- exchange of information between scientific-production organizations and institutes of St. Petersburg and the access to international, scientific-technical data bases and banks;
- joint expertise of innovation projects;
- installation of telecommunication facilities for the efficient interaction between scientific-production organizations and institutes of St. Petersburg;
- co-ordination of international relations, joint participation in international scientific-production and innovation projects;
- assistance in the training of the research personnel losing jobs as a result of the defence conversion.

In conclusion, I would like to underline that the problems in the scientific-technical sphere of our industry are related both with the functioning of science itself and the general economic situation in the country. The particularities of the transition period result in the non-standard Russian conditions. Therefore, to solve the problems, one should choose the ways specific for Russia. It is important to constantly analyze the tendencies in this sphere, to summarize the experience of the efficient reorganization of industrial enterprises and to help to introduce this experience into practice.

MAJOR FACTORS OF SUCCESS IN SMALL BUSINESS SPHERE

Alexander KOBYSHEV
"Progress" Management and Marketing Centre

Small entrepreneurship based on private initiative has been practically forbidden in the USSR since the dramatic end of Lenin's "new economic policy" by 1930's. It was only in 1989 when the new law "On co-operatives" partially legalized private entrepreneurship in the Soviet Union.

The break-up of the USSR in 1991, privatization of state enterprises in Russian Federation and certain legislative steps were followed up by tremendous growth of the number of small enterprises (SE) in Russia.

By 1995 the total number of SE in Russia was estimated at the level of approximately 800 thousand[1] (10 % GDP, 13 % of total employment). However, the economic and legislative instability accompanied by the absence of practical state support, was followed up by 7 % decrease of the number of SE in 1995[2].

The success of SE is usually seen in Russia as being closely correlated with mostly external environment. The latter is blamed for setting up obstacles of political, legislative, economical and social character both on federal and local levels. As reported by Russian association of small entrepreneurship development, out of 35 basic problems faced by SE, 29 are directly connected with external environment[3].

Two major of those are punishable taxation and incomplete legislation. It is enough to mention that the definition of small entrepreneurship has been formulated officially only in June of 1995 in the Federal Law "On State Support of Small Entrepreneurship in Russian Federation".

On such a background the internal problems of SE in Russia were hardly considered as of importance. However, a closer study of these bring into focus fundamental aspects explaining success of some or failure of other small businesses operating in the same external environment. These aspects may be broadly subdivided into three groups:

1. General business culture;
2. Psychological aspects;
3. Business education.

Decades of planned economy, through which private entrepreneurship has been condemned as criminal activity, have atrophied business culture. "Proper" management

---

[1] Official statistics on registered SE.
[2] Analytical report "*Development of small business in Russia*". International Conference "*Small business in Russia*". St.Petersburg, 1994.
[3] Ibid.

C.M.R. Verkoeyen et al. (eds.), Management Training in High-Tech and R&D, 17–21.

has been reduced to fulfilling (preferably overfulfiling) of state plan and proper and in-time reporting. For directors of state enterprises "proper" management also ment shaking out of corresponding ministry solid budget, ensuring stable supplies and equipment and, by the end of the year, correcting state plan.

"Unimportant" matter of cost effectiveness has been considered irrelevant, especially in military oriented industries (about 70 % of St. Petersburg industry by the end of 80's). Unproportionally large inventory has been considered as normal. Human resources management has been delegated mostly to retired military heading personnel departments. Such were just some realities of the recent past directly influencing the present state of business culture, which may be summarized as at table 1.

Observation formulated in table 1 are directly related to psychological swing which is being witnessed in Russia in 90's. This swing may be considered as reflection of initial stage of economic transformation in direction of market economy. An attempt to formulate business-related psychological approaches is presented in table 2.

Contents of tables 1 and 2 make quite clear the basis of SE internal problems. They also make evident a proper solution which lies in the field of business education.

According to recent research[4] conducted at St. Petersburg (7 % of total number of SE and 3.5 % of population of Russian Federation) most of entrepreneurs of SE have solid professional education (tables 3, 4). However the majority of them belong to age groups, which are specifically effected by recent economical past (table 5). About 77 % of entrepreneurs had previous managerial experience at large state enterprises, including 36 % at junior management level (up to 20 subordinates) and 25 % at middle management level (20 to 99 subordinates).

Latest comparison[5] of top 10 managerial qualities arranged in priority order for Russia and Western countries reflect the present state of entrepreneural culture difference (table 6).

By our estimation about 50 % of entrepreneurs attended recently some seminars, mostly short term, devoted to different aspects of entrepreneurship. Such short-term (2–3 days) training can not be condisidered satisfactory both in term of deapness and, often, contents of programmes.

Compared to average managers of SE entrepreneurs evolving of retired army officers have sufficient benefits. The army training allows them to avoid internal problems at the start of SE, which, combined with the high organizational level secures stable development of SE at complicated external conditions. However even this group of entrepreneurs need additional intensive business training.

A variety of business training programmes is being offered at local education market. This market is yet another problem due to the lack of qualified local trainers and materials.

Objective problems of executive training include constant deficite of time and financial resources. The lack of specialized intensive training programmes developed for SE managers is also evident, while demand is growing, though rather slowly.

---

[4] Analytical report "*Development of small business in Russia*". International Conference "*Small business in Russia*". St.Petersburg, 1994.

[5] Financial Izvestia, № 86(320), 05.09.1996.

Table 1

Some aspects of business culture and managerial practice

Aspects	As accepted by Western entrepreneurs	As understood and practiced by Russian entrepreneurs
Social responsibility	Company success plus benefits for society	Company/personal success by all means
Business obligations	To be respected	Considered according to current situation
Personnel	Vital component of success	Employees, technical performers
Marketing	Major instrument for satisfying customer's needs	Selling off products available
Strategy	Corner stone	Everyday tactics for survival
Financial management	Production influencing factor	Production reflective factor
Accounting	Strict	"Creative"
Profits	Stable, reasonable	Today, maximum (minimum reported)
Taxes	To be paid	To be avoided
Communication	Open vertical / horizontal	Closed vertical

Table 2

Some business-related psyhological approaches

Internal SE probelms	As treated by Western entrepreneurs	As treated by Russian entrepreneurs
Psychology	Synergetic effect to be achieved	Due winner
Competition	Challenge, possibilities	Game with no rules
Risks	To be evaluated	To be avoided
External share holders	Co-owners	Enemy
Customers	Customer focus	They are many, we are few
Conflicts	Creative energy to be managed	Destructive energy to be avoided
Decision making in crisis	By consensus	Personal

Table 3

Entrepreneurs of SE. Professional education. St. Petersburg, 1993

Field of activity	%
Industry	62
Construction	60
Consultancy (law and economy)	50
Medicine	48
Trade	40

Table 4

Entrepreneurs of SE. Education level. St. Petersburg, 1995

Education level	%
Higher*	78.5
Special secondary	16
General secondary	5

* Including 7 % of academic degrees holders.

Table 5

Entrepreneurs of SE. Age groups. St. Petersburg, 1993

Age (years)	%
Below 30	8
30	34
40	37
50	16
Above 60	5

Table 6

Top managerial qualities, arranged in priority order

№	Russia	Western countries
1	Personal achievements	Leadership
2	Personal connections	Managerial effectiveness
3	Loyalty	Perspective vision
4	Entrepreneurship	Agreement attaining
5	Leadership	Entrepreneurship
6	Analytical capability	Role flexibility
7	Problems solving	Receptiveness
8	Managerial effectiveness	Flexible style
9	Perceptiveness	Analytical capability
10	Receptiveness	Problems solving

PARTNER-SEARCH AND NETWORKING

Carolus VERKOEYEN
TNP Management, Belgium

1. Introduction

When discussing the conditions for a successful transformation of organizations and enterprises, including research and development institutes, from a planned economy system to (e.g.) a market economy system, the ultimate success is dependent on that element, that was not solved or badly solved: a so called critical cucces factor.

It is not always possible to predict which one will be that weakest element; however, it is clear that on many occasions the attention is focused on what is nearest to the attention of direct responsible staff, management and policy makers and for that reason seems to be most important (the problem of selective perception).

In this way they may overlook some of the elements that are really undermining the attempts for a successful transformation.

During the Advanced Research Workshop in Saint Petersburg many aspects have been highlighted: aspects that were concrete as they are familiar from the past, or from today's awareness and reality, as they are experienced; the role and responsibilities of management, skills of a manager, etc.

It should not be forgotten that process Russia and other newly formed states are going through now is a reversed process from the state guided transformation, as the Soviet Union experienced after 1917.

It is not unusual that some aspects which will be important in the (nearest) future are forgotten or do not get the attention that they need. One reason, is that Russia then was an agricultural state and another reason consists in the fact that since 1917 new phenomena have occurred in capitalist society, like marketing.

In this presentation one of those elements will be highlighted: the future, wanted, desired partner; an element of marketing, as will be argued in this article.

2. The partner

The future partner of an R&D-institute may have different profiles, and in practice such an institute will have several different types of partners:

- other research institutes;
- (industrial) producers, buyers of knowledge;
- financiers;
- institutes for testing, certifying, the protection of intellectual property, etc.;
- authorities.

C.M.R. Verkoeyen et al. (eds.), Management Training in High-Tech and R&D, 23–37.

Analysis of such a list leads to the following observations:

- from the functional point of view these partners fulfill different, mutually complementary roles;
- the partners are of relevance at different stages of R&D "production" cycle ;
- all of them are crucial to secure the effectiveness of an R&D institute, now and for the future.

This presentation will deal with an element, that – by the experience of the author – is very often neglected and/or underexposed:

– the search and the identification of the proper partner(s).

3. Why a partner-search?

The most simple answer to this question would be: "because usually it is a problem". However, such an answer is not very helpful to achieve better results.

For that reason the following critical success factors are distinguished:

- To have a clear and adequate image of an intended partner; it should be possible to answer the question: "do we need a partner and do we have a clear picture of what we need from our partner?".
- The willingness of the potential partner; do we have an answer to the question why that partner would be able and willing to fulfill these needs? And at what costs/price?
- To be of value to the intended partner; do we have a value to our partner? Is that understood by that partner or can we make him understand?

At the initial stage, when we start thinking about a partner, discovering the discrepancy between our expectations and reality is very often most disappointing.

It is exactly for this type of reason that a partner-search is an activity that needs to be planned very carefully and executed accordingly.

On one hand not to disappoint ourselves, but, – on the other hand – also not to disappoint our potential partner.

4. Preparing a plan of approach

The elements of a well prepared partner-search include the results of a systematic analysis of a company's own strengths/weaknesses and needs/offers towards a future partner, as well as a time path.

For the systematic analysis an R&D institute may be considered as any other company that is producing or rendering services.

And – in consequence – all the elements, playing a role in any business plan have to be considered:

- the ultimate aim of a company (on strategic level);
- the way the different production means are organized (tactical level);
- how the day – to-day work is organized (operational level).

From here more detailed matters have to be clarified, like:

- what profile of workers does the institute have and what is needed; the number of people needed;
- how does the market look like, where is the market, what does the market want or need, what is the market willing to pay, who/where are the competitors;
- the planning of resources; apart from people, finances;
- prognoses regarding investment and return on investment;
- etc.

Such a process of making a business plan is demanding time and energy, and the following problems have to be faced:

- management of R&D institutes in Central and Eastern Europe are not used to work with this type of business plans (note: the type of planning, used under the circumstances of the plan economy system has nothing in common with a realistic business plan);
- going through the process of making a reliable business plan may lead to shocks, as confrontations between the self-image of an R&D institute and the way the market sees the R&D institute may be quite disappointing;
- under present circumstances R&D-institutes in e.g. Russia have to solve problems which have never existed, or existed but have never been faced by R&D-staff itself. In particular one has to think about the next element in the production-chain: who is going to implement the findings of R&D work.

With these last questions the nucleus of this contribution is touched, and this can be summarized in the following statements:

1. The traditional staff of an R&D institute is not prepared for promoting it's own institute and results.
2. R&D institutes need a marketing strategy, including methods, techniques and tools.

The second statement is the most crucial, but neglecting the first one would imply neglecting a sure critical success factor.

5. The staff of an R&D institute

This contribution will not deal with the question weather an R&D-institute should have a rather flat structure, a matrix organization, etc.

In the context of this contribution attention is attracted to the fact that by experience working-teams tend to consist of four types of persons:

- promoters/initiators;
- supporters;
- analyzers;
- decision-makers.

Each type is based on characteristics that can be measured by certain tests (two in particular).

As a result of scores in those tests a person is positioned in one of four quadrants, where each quadrant is labelled as mentioned before.

As this score is the result of a combination of two tests a person may score in an extreme corner, or close to the cross of the axes.

This testing technique is not the subject of this contribution, but the relevance of it may be proved by the fact, that in Western Europe these tests are very often used to check if e.g. project-teams for building information-systems are well balanced.

In case of project-teams there is a direct possibility to influence the composition of a group; in case of a standing organization such a possibility is limited.

However, this way of considering an organization simplifies the understanding why an organization fails to achive the aims and meet expectations: expectations of the outside world, but also expectations from inside.

A global scanning of many R&D institutes confirms that analyzed types are well represented in these institutes, if not overrepresented. Usually, the supporting type is also sufficiently represented. The decision-making type may appear to be the first problem, as the managers of these institutes are very often researchers themselves (so more is the analyzing type).

Here a parallel can be observed with other professional organizations, like e.q. hospitals. The director of a hospital used to be a medical doctor, with great achievements as a medical specialist, but without being sure if he is a good manager.

Over the past twenty years, Western Europe showed a tendency towards hospital managers, who didn't have a medical qualification themselves, but who had been trained to become managers in a medical environment. It was an improvement in several respects.

The last category mentioned is the promoter. The promoter has two main characteristics:

- he/she has "ideas";
- he/she is able to enthusiast others.

The first element has to be fostered and what is useful has to be adapted for the organization, the second element can be exploited for the benefit of the organization.

6. The promoter

The "ideas" of a typical promoter are usually coming fast, and are followed by new ones. This may result in tensions in the organization especially as the analyzing type of colleague will take his time to process the presented ideas. And not only that, after some time he will come with questions, with remarks, with pluses and minuses, with advantages and disadvantages, etc.

As the promoter at that moment will already be dealing with new ideas, this situation will cause tensions.

It will be very beneficial for both the individuals and the entire organization, if this mutual complementary value will be recognized.

The enthusiasm that promoters show is a characteristic that is useful for the organization, provided this enthusiasm is steered in such a way that it fully contributes to achieving the aims.

Anyhow, the function of promoting the institute, its work, its results, etc., has to be taken care of. It is not an aspect that will be settled just by itself and the institute here even has a choice; the choice of having this promotion power inside or outside its own organization. Both options have advantages and disadvantages.

To name a few:

– promotion, public relations and marketing require professional skills, training and mentality. Promotion is a specialized job that cannot be performed as just an extra task, as something you can be doing during a few hours a week. So, if the institute is too small to appoint one person with this task especially, it can delegate the task (sub-contract it) to a specialized company. In this case even a small R&D institute can develop a very professional marketing (P.R. / promotion) policy and activities.

– P.R., promotion and marketing are still disciplines, fields of expertise on their own. Thus subcontracting opens the possibility to subcontract these respective tasks to different, specialized companies.

– of course this sub-contracting of work also has a vulnerable aspect: the contracted companies have to be briefed very well. Such a briefing is absolutely necessary in order to make the contracted companies understand what mission statement you want to communicate, what your strategic aims are and what the profile of your company is, as you want to advertise it.

Of course real professional companies in P.R., in marketing and promotion do understand the relevance of a proper briefing, but in Central and Eastern Europe too many companies consider themselves to be specialized in these field, without really being.

Evidently the advantage of bigger institutes capable of having such a department for P.R., and/or promotion within their own structures consists in the fact that communication about these aspects is limited to internal communication; quite a convenient situation.

7. From having a promoter to promotion, P.R. and marketing

It should be underlined that management has the responsibility to take care that a marketing (and P.R. and promotion) strategy is established. But this responsibility basically means that they have specialists preparing and doing the job, based on management's decisions.

A second, very important, task managers have is to perform at certain events, places and occasions. They are very often the face of an institute. Again here, the P.R., the marketing and promotion strategy may prescribe where to act, when, how far, whom etc. In short all basic questions of communication have to be answered.

The why question seems easiest to answer; even in the situation it may be a problem to concretize a marketing strategy on tactical and operational level. The same with a proper P.R. and promotion strategy.

However, when concentrating on the *to whom* question and also on how and *where* questions, it will be noticed what aspects in the why question have to be detailed or to be elaborated.

8. Marketing towards whom

The question to be answered is: who is relevant for an R&D institute. When answering this question of course automatically the why question will pop up again.

Within this contribution no full and complete list will be presented or even can be thought of. Each single institute will have to answer this question by itself. But a first possible step is the choice of a framework to come to an inventory.

R&D institutes have "clients".

So it should be answered who are asking for R&D, who are ordering R&D or who are using the results of R&D, even without asking before.

Directly related questions are:

- where are these clients;
- what relation does the institute have with them, or if this relation doesn't exist;
- how can such a relation be established.

As a consequence the following questions have to be answered:

- how can the relations be maintained, by what means;
- the frequency of the contacts in such a relation;
- particular forms; some acceptable in one relation, but not in another one;
- how big are the respective target groups;
- how relevant.

Questions as put before should fit within a schedule of functionalities, both towards R&D institutes and their relevant environment. Which methodology is chosen to determine these functionalities is open for discussion, but the choice of a methodology and any other analytical framework will assure the completeness, that otherwise may not be achieved.

One of the frameworks that may be considered is the framework of the production column/chain.

The production column describes the relations between enterprises in the whole process of production, starting with raw materials and finishing with the end product delivered to the end user. Using such a production column for analyzing (potential) relations relevant for an R&D institute is very illustrative, as may be understood from a comparison between the following two schemes (Fig.1).

This way of working also offers an R&D institute the possibility of determining and distinguishing "sure" R&D and potential R&D.

Schedule 1: The classical production column/chain.

Schedule 2: The Production column/ /chain, the opportunities for added value by/through R&D.

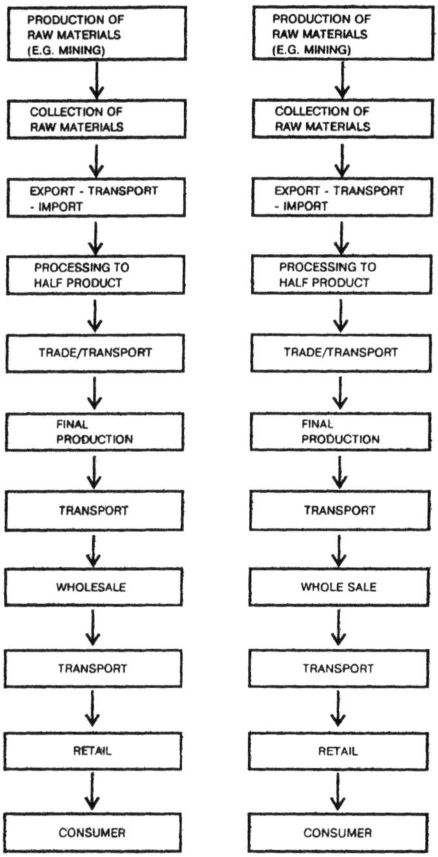

<u>Possibilities for R&D</u>
- environmental protection
- improved mining techniques
- improving storage, (un)loading and transport
- developing systems for logistics management
- transport monitoring systems
- developing more economic production techniques
- energy saving
- product innovation
- trade information systems
- product innovation
- marketing and business planniag techn. systems for JIT
- multi-modal transport
- improved transport and export packaging
- techniques and tools for internal transport
- developing adequate consumer packaging
- stock and sales information systems; implementing international barcode

Fig. 1

And this touches upon the essence of the marketing of an R&D institute. The result of the approach, as described in previous section, is an overview of actors to monitor, to be in contact with and – in some cases – to influence.

As a matter of fact the analysis, will also result in detecting indirect actors.

E.g. if one finding is that the processing of raw materials should be done less polluting, it may not be the direct interest of the producing company as it may increase production costs.

However, it may be a reason to discuss the needs for clean production processes with authorities, who may give a legal framework to this requirement. Sometimes it is a matter of using existing rules for organizing an awareness process towards such a polluting industry. E.g. within the European Union more and more regulations are accepted not allowing certain products on the European market, if the production process or what remains after using or consuming the products is too polluting.

As it is not environmental protection which is the subject here, but the way of thinking that R&D institutes – in order to create a market for themselves – have to implement, this example will not be elaborated any further here. The way of thinking and the type of approach is important for R&D.

It is possible to analyze the production and consumption environment from the point of view of:

- safety and protection;
- medical aspects;
- packaging and transport;
- financial control and logistics;
- etc.

So, gradually a picture arises where the environment of an R&D institute is a highly segmented environment, including:

- institutes directly involved in production, transport and trade;
- institutes indirectly involved like (semi)governmental institutions, financing and insurance, certifying institutes and institutes for norms and standards, etc.;
- but within most of these categories, people are involved at different levels:
  - strategic;
  - tactical;
  - operational.

In other words, a matrix has to be developed, allowing an R&D institute to identify blanc spots and to strive for a certain completeness. Anyhow the result of this exercise is a pretty complete overview of target-groups for marketing, P.R. and promotion, and .... awareness within the R&D environment, which entails the next problem to be solved: how to organize the communication with all these (sub)groups and categories of target groups.

At least the following elements can be identified which can help decide who – on behalf of an R&D institute – should be involved:

– first of all the discipline is the subject (of marketing, P.R., promotion); apart from the research legal, financial-economic and general policy are the fields which mostly have to be covered;

– the second consideration regards the level of the actors: strategic, tactical or operational, and this is closely related to the third element:

– the size of the target group.

If the target group consists of the Minister of Environmental Protection, it may be the case that the President of the company is in charge. If the target-group indicates all users of detergents in the country it may be considered to make the department of external relations responsible.

9. Networking

By now the need of directing and conducting this process may be understood. The number of potential actors in the communication process between an R&D institute and its relevant environment may become very complicated; if involves many actors and is multi-level.

The need for coordination now arises from the following threats:

(1) It has to be secured, that the contents of communication between an institute and its relevant environment is consistent, independent on who is communicating on behalf of the institute or with whom.

(2) As far as communication costs, both time and money are concerned, not all marketing, P.R. and/or promotion activies which are theoretically possible will be organized. So choices have to be made and priorities formulated.

The evident way consists in appointing a person or department, responsible for coordinating this communication.

Such a coordination then has to deal with contents (the message), forms, structures and with processes.

The <u>contents</u> to be communicated, information to be collected, etc., is primarily a matter of strategic and tactical levels in the organization. They have a clear responsibility in steering an organization. Marketing specialists may assist them in this job and even have to execute the job. But ultimately it is the responsibility of management to decide what is communicated to whom and why. About the when and in what form the specialists in communication may have a strong say.

The <u>form</u> includes aspects like house-style, but also the choice of media and events: from press releases, (own) periodicals to the participation in conferences and exhibitions. Some examples will shortly be described in the last paragraph of this contribution. A person or department (or a sub-contractor) may be expected to be an expert in developing a house style, choosing or advising forms of communication within the respective areas of marketing, P.R. and promotion.

The <u>structures</u> through which communication is or will be organized are often existing ones, but not always. Decisions have to be made on what part of these structures is useful for the marketing, P.R. and promotion aims of the company. If none is, what can be done about it: choosing alternative structures, adapting the existing ones, or whatever else possible. Anyhow analyzing and being aware of the best and worst ways for structures communication is very important. Also one should not forget about the internal structures of the organization. P.R. is as much an internal activity to the organization as it is an external one. Communicating the aim of an organization, its mission statement, to its own employees is a condition that has to be fulfilled if the organization wants to achieve proper, tuned communication with the environment, the external structures.

One aspect about structures that should not be forgotten is the existence of informal structures. Informal structures are more or less stable relations and communication between persons, which are not necessarily reflected by the formal positions these persons have in the actual existing formal organizational structures. These informal structures play an important role especially in post-communist Europe. But not only there, this is the case also in Western Europe and North America.

As to the processes it has to be understood that processes are always interrelated to the structures within which they take place. However, by focusing on the processes it is possible to focus on the social-psychological aspect of the marketing-communication, P.R., and promotion. It is the task of the specialists in communication to observe both individual processes and collective ones.

The individual processes are most relevant when considering the effectiveness or the lack of effectiveness within the organization itself. As far as external communication involves small target groups, key-players (opinion leaders) in the environment of an institute are important. The particular processes which have to be observed are e.g.: obstruction (and after answering the why question, how the problem can be solved) and ignoring. Ignoring may be a matter of lack of awareness, a lack of involvement or acting in sufficiently consistency to what was expected by the architect of the communication (this last aspect may be caused by the lack of understanding of an institute's aims relevance and the choice of wrong instruments, timing, etc., for the communication).

As to the collective processes, the character of possible misfits in the communication has to be understood. Is it a matter of collective concern (e.g. about the pollution of the nearby factory) or is it a matter of opinion leaders who effectively are able to mobilize the collective opinion (e.g. about the pollution, caused by leaking oil-pipes in Siberia) .

Apart from understanding the character of these collective processes, the tools used by "the others" have to be analyzed. Such mass media as television and newspapers fulfill an important role in this, but not only they. Also mass-manifestation, the lobby to a parliament etc. can fulfill such a role. And as a consequence an institute – if it wants to correct certain parts of not appreciated elements in the communication may choose the same or even stronger communication aims.

10. The networking atlas

So far no real, existing networks have been described. And it will not be done in this paragraph. By definition the network of each institute is different. A common element is the fact that the work, as may be understood from this contribution so far, will lead to concrete names, addresses, categories of target groups, etc.

Describing these names, addresses and categories, sorting and ordering them, deciding how they will be approached, where, when, in what way is an important process that will enable an institute to develop a real marketing policy (and P.R. and promotion).

It can be extremely helpful to organize all this knowledge on target groups, internal structures and external structures, in a document: an information and/or networking atlas. Such a document enables conscious decisions instead of random ones or non-decisions at all.

In particular such an atlas is an instrument to form target groups at levels: micro, meso and macro. This distinction is of importance in two ways.

(1) according to the dimension of how nearby the target group is; the nearer, the more relevant; the farther away the less relevant. To avoid a possible misunderstanding this being nearby can be understood as being nearby geographically, but also mentally or by real power (like the power of certain institutes in centralized states). The problem of institutions in post-communist Europe is that they have to deal with both: still central Government, Ministries which have power when it comes to the budget, to the distribution of foreign technical assistance. On the other hand, the bureaucrats next door having power to make life miserable or to ease it up.

(2) as a consequence of the first element the aspect of possibility (and sometimes also the necessity) to influence has to be recognized. Also here such an exercise as distinguishing groups and people in belonging to the macro, meso or micro level is useful, as it will help to determine which communication instruments have to be chosen.

After all remarks, offering possibilities to be reflected on the networking of a company or an institute, it still should be reminded that the most important element in all is people. People are communicating, people are sending messages, people are intended to receive and understand these messages; to decode them and to fit those messages in their reference framework. All methodologies, all instruments and tools are intended to support and facilitate this work. But people communicate with people and people meet people.

This leads to the final paragraph: the description of an instrument that until recently was differently perceived in post-communist countries-fairs and exhibitions as a place where people meet people.

11. Where people meet people

In the past fairs and exhibitions in Central and Eastern Europe were dominantly:

- either big consumer fairs, with hundreds of thousands of visitors;
- or trade fairs, where contracts were concluded and signed.

The success of the event could be measured according to the number of visitors or the number of contracts and the volume of the contracts concluded.

In Western Europe these big consumer fairs hardly exist any more, or – if so – on specialized subjects: automobiles, bicycles, holidays.

But also for all kinds of branches, economic activities, specialized fairs became more and more important; big and important fairs have been developed in agriculture, packaging, medical equipment, environmental technology, building and construction, etc.

Almost all visitors of these exhibitions are specialists. As a consequence it may be understood that it is no longer the absolute number of visitors which is decisive to conclude if a fair was successful, but the relative number of visitors. In the Netherlands a furniture fair is successful if 15,000 people visit the event. Because in that case almost 100 % of the people working in the branch have been there. However, a building and construction fair is successful when over 100,000 people were there: 80 %.

Another remarkable aspect is that concluding of contracts the is no longer a central topic. Only 3 % of the exhibitors expect to conclude contracts. For most of them it is a matter of maintaining existing relations, establishing new relations and in some cases of presenting new products or services.

Another observation is the tendency towards the move, not only from horizontal (broad) fairs to fairs within a specialized field, a segment, but also to more vertical fairs. That means an integration of all aspects from raw material to end product. A fair (e.g.) on intensive animal breeding, includes everything from the genetic material to the piece of meat on the plate. In that case the possibilities and opportunities for R&D are just endless.

In order to get a clearer picture about this type of events and the potentials they represent, this article is finished with some data on recent events. The events have been chosen as in one way or another they may open opportunities for the participants in the Saint Petersburg R&D workshop.

(1) ECOTECH '95 (December 5–8, 1995)
The exhibition range included products and services for:

- collection, transport and processing of non-chemical and chemical waste;
- soil decontamination and protection;
- air purification;
- sound pollution control;
- recycling;
- waste water treatment.

Next event to be scheduled for.
(2) VAT '96 /European Subcontracting Trade Fair (January 30–February 2, 1996)
The exhibition range included the following segments:

- main suppliers, mechatronics and special machine construction;
- metal machining;
- plastic and rubber processing;
- surface technology;
- tools, stamps, matrices and fine mechanics;
- electronics and electrotechnology;
- raw materials and semi-manufactures;

- technical glass and ceramics;
- engineering and product development;
- consultancy and information.

Next event to be scheduled for.

(3) DIS '96/ trade fair and conference for information, image and document processing systems (February 20–23, 1996)

The exhibition range comprised the following categories:

- services;
- trade literature;
- technology;
- hardware;
- software;
- micrography;
- technical drawing management;
- multimedia;
- optical media;
- text/information conversion/recording;
- communication;
- filing systems;

Next, event to be scheduled for.

(4) TECHNI–SHOW '96 Trade Fair for Industrial Production Technologies (March 18–22, 1996)

The exhibition range comprised the following:

- measuring technology
- compressed air and vacuum technology
- machine tools
- synthetics processing machinery
- linkage technology
- industrial automation and computerization
- production automation and computerization
- workshop technology
- equipment for non-cutting metal machining
- equipment for metal machining
- surface treatment equipment
- information and knowledge transfer.

Next event to be scheduled for:	ECOTECH	VAT'96	DIS'96	TECHNI-SHOW'96
Net exhibition area/m^2	17,018	15,530	3,970	43,064
Nr. of direct exhibitors	347	413	96	565
– from outside Holland	33	51	–	48
Nr. of visitors (total)	27,080	12,874	7,605	65,238
– from outside Holland	1,916	1,195	139	3,766
Main sector of visitors:				
– Metal products industry		33.8 %		77.2 %
– Public admin./publ.serv.			22.5 %	
– Industry	19.7 %		12.6 %	
– Services/trade in real estate			14.8 %	
– Computer service and information techn.			14.7 %	
– Environmental	14.4 %			
– Local public sector	14.1 %			
– Machine industry		12.6 %		
– Electrotechnical ind.		8.2 %		

These data are published in the reflex-areas. Apart from quantitative facts also qualitative data are published. A selection:

On ECOTECH: Most visitors are interested in chemical waste, soil and waste water. One of five visitors was a decision maker, also one of five was employee of a technical department. The most important reason for visiting was to investigate new products (43.9 %), the second reason was to investigate market developments (27.9 %).

VAT was visited by the representatives of very small enterprises (1–5 people; 17.4 %), but other 15.9 % of visitors came from enterprises having from 51 to 100 workers. However, 6.7 % of the visitors represented enterprises with more than 1000 people. Fields of the main interest were metal machining (50.0 %), plastic and rubber processing (34.2 %) and raw materials and semi manufactures (20.6 %) (multiple answers possible). For most visitors the main aim for visiting the fair was to keep abreast of new technological developments (35.2 %). Another important aim was to establish new contacts (19.4 %) and/or to maintain existing relations (14.0 %). Only 1.3 % came to place orders at the fair.

In DIS board management (13.1 %), heads of automation (9.8 %) and project managers (7.7 %) were mainly interested. They were most interested in filing systems (38.3 %) and software (37.7 %). Text/information conversion/recording followed at the third place (24.0 %). It was remarkable that 31.9 % of the visitors came from enterprises having over 500 employees.

At TECHNI SHOW 38.3 % of the visitors was interested in processing equipment for metal machining, followed by interest to processing equipment for plate metal work (16.1 %) and machine tools (15.7 %).

Immediately after production specialists (25.3 %) most visitors were decision-makers (23.5 %). Their most important reason for visiting the event was to keep abreast of new technological developments (48.2 %).

The last interesting element that should be mentioned is the following: most visitors are coming to the event on invitation of one of the exhibitors:

ECOTECH '95	60.3 %
DIS '96	56.8 %

The consequence of this fact is that an R&D institute participating in such an event will meet not only other participants as possible future clients/partners, but also the clients of those other exhibitors. This is a networking instrument in now-a-days industrial environment, where hardly compatible alternatives are available.

The networking already existed under the plan economical system. Then it was to secure privileges. Under post-communist system the networking is initially used in the same way. Now it is the task to upgrade the networking according to the rules of the market, which means a development of proper marketing, P.R. and promotion. Hopefully this contribution may fulfill a small role in that process of upgrading.

PROBLEMS IN R&D MANAGEMENT TRAINING

Vladimir IVANOV
The Ministry of S&T Policy of the Russian Federation,
Yekaterina BESSARABOVA
The Analytical Centre of Science and Industry Policy

Transition to market economy has entailed a number of changes in all spheres of economy and, first of all, in production and research fields. The main trends typical of large enterprises have become as follows: changes in the forms of ownership, setting-up of holdings, establishing financial-industrial groups, as well as breaking-up of large enterprises with the purpose of establishing a number of smaller firms on their basis.

It has become a distinctive feature that the number of small and medium enterprises (SME) in the field of "science and scientific services" is increasing, which is brought about by the disintegration of large research and production companies, research institutes and manufacturing enterprises. Though the importance and significance of this-kind enterprises are proved by the world practice, in Russia they face a number of problems, among which the following should be mentioned:

- difficulties with the search for a potential investor;
- absence of current assets;
- high price and instability of rent conditions;
- official and unofficial expenses of ensuring the safety of both a company's property and its activities;
- need in costly technological and measuring equipment;
- restricted access to scientific and marketing data bases;
- need in preparing business-plans of projects;
- necessity of training managers for small enterprises.

The above problems may be divided in two groups. The first one unites financial and economic problems which may be explained by the general instability of the economic and political situation in Russia, imperfection of the Law and absence of a clear mechanism for investments in scientific sphere. The second group of problems is connected with the necessity of establishing centers for the support of new technologies and innovations capable of assisting SME's in settlement of their current problems (legal and economic consulting, search for investors, ensuring access to information required), as well as in providing the managers and employees of these enterprises with an opportunity to take necessary training courses.

C.M.R. Verkoeyen et al. (eds.), Management Training in High-Tech and R&D, 39–40.

Today there exist a number of institutions providing an opportunity of getting economic and management training:

- governmental and non-governmental educational institutions mainly oriented to students training, but also rendering chargeable educational services, so that people having experience of practical work could get the second higher education (economic or legal).
- Specialized qualification-improvement institutes for specialists of certain fields.
- Non-governmental educational institutions established at associations, foundations and other structures oriented to providing financial support for small and medium businesses.

Obtaining the second higher education, for which it takes not less than two years, proves to be impossible for SME managers and employees, as they cannot leave their enterprises for a long time. Training in specialized institutes of qualification improvement is of no help either in solving concrete problems the main among which are entering market with products (R&D's) already available and search for a foreign partner for joint projects implementation. Therefore, non-governmental training institutions established at organizations oriented to ensuring financial support for small and medium businesses in R&D field appear to be the type of educational institutions most acceptable for SME's. Such training institutions have a course training system including basic training courses (2–4 weeks) and primary presentation of a project (drawing-up of a business-plan included) in Russia with further practical training abroad (2 weeks) at enterprises similar to those where the trainees work in order to improve their projects and further tender them for getting funding.

As a rule, course training systems are chargeable for trainees and are also financially supported within the framework of different international projects.

The following courses are distinguished as the most significant ones within the framework of training programs:

– Russian and foreign law, especially in intellectual property;
– currency and customs regulations;
– management and marketing of high-tech products;
– foreign language;
– business communication skills;
– disciplines connected with mastering new informational technologies.

Branch orientation of training courses ensured by their “attachment” to employees' practical problems proves to be a very important moment.

MANAGEMENT SKILLS FOR THE SCIENCE BUSINESS

Hanns Peter MUTH
Enterprise PLUS Ltd., United Kingdom

1. Introduction – Definitions

The theme of this Workshop refers to "High-Tech Industries, Science, and Research and Development." This broad definition includes scientific activities with a definite commercial purpose (e.g., "high-tech" industries and R&D aiming at developing marketable products or processes) and others that may or may not have a commercial purpose (e.g., theoretical or "pure" science and non-directed research).

The term science business as used in this presentation includes only scientific endeavour with a commercial purpose and covers all activities – whether they are "scientific" or not – designed to achieve that purpose.

Managers – in the science as well as any other business – are judged by their effectiveness and efficiency. In casual conversation these terms are often used interchangeably, but they most definitely are not. The term effectiveness describes the success of an activity in terms of the achievement of its objective or purpose; the term efficiency refers to the economy of resource utilisation.

Finally, I want to introduce the somewhat simplistic distinction that the science business requires competent managers if it is to be run efficiently and effectively, whereas scientific endeavours without a commercial purpose call for able administrators. I shall be talking about the training of managers, not of administrators.

2. Two main themes

My presentation has two main themes:

(1) First, managers in the Science Business must be managers first and scientists second.

(2) Secondly, the skills required for management in the "Science Business" do not differ from those required in any other business.

2.1. THE "SELF-IMAGE" OF MANAGERS IN THE SCIENCE BUSINESS

When dealing with managers in the science business I have sometimes observed an inclination to see themselves as "scientists first, managers second". This applies

C.M.R. Verkoeyen et al. (eds.), Management Training in High-Tech and R&D, 41–50.

particularly in countries such as Russia, where being a scientist has, for most of the country's history, been a more certain way to the top of the social pyramid than being an entrepreneur or manager.

Given the gap in public esteem for science and business, respectively, it is tempting for scientists who are also managers to set themselves off from the pack of under-appreciated businessmen and encourage the development of an élitist sub-group in the management profession: that of the "scientist as manager".

This may indeed be tempting, but it is also highly dangerous, because it is apt to encourage the scientist as manager to look down from his Olympus on the "manager only". Membership in the élite club of scientist-managers can then be used to shield members from being forced to compete with "mere managers" who, while better managers, nonetheless are inferior because they are not also scientists.

To be at their best in both worlds, managers who are scientists must see themselves as "members in good standing of two different clubs": Science and Business, each with its own conditions of membership, rules of professional conduct and criteria for measuring success.

As long as they are engaged in managing a business, their membership in the business club has absolute priority: They must be managers first and scientists second. In the scientists' club they may be scientists first and managers second. For all I know, they may also belong to a sailing, bridge or a mountaineering club and, to be fully accepted, may have to be sailors (or bridge players or mountaineers) first and then scientists or managers.

2.2. THE SKILLS REQUIRED BY THE MANAGER WHO IS ALSO A SCIENTIST

As a member in good standing of the business club, the scientist-manager plays by the rules of that club. It is immaterial that he or she also knows the rules of the scientists' club, or that of the golf, tennis or sailing club.

To be able to play by those rules, the member will benefit from being conversant with the skills that make a good manager, i.e. from being a trained manager.

That, finally brings us to the main topic of this presentation: What skills must a manager possess to rise to eminence in the business club?

3. Management training for the scientist-manager

The simple answer to that question is, of course, that they are the same skills any other manager must possess. But what are those skills?

To manage is to Plan, Implement and Control the Effective and Efficient Use of Resources in the Pursuit of Clearly Defined, Realistic Objectives. Let us take a closer look at each element of this definition.

3.1. SETTING OBJECTIVES

Let us recall the earlier definition of the science business as scientific endeavour with a commercial purpose. Precisely. The scientist-manager, and any other manager, must have commercial objectives that are realistic and clearly defined in terms that allow measuring their achievement. The manager's training must enable him/her to define and set such objectives.

Objectives must be commercial:

- "To make the world's best laser diodes" is not a commercial objective and therefore not an objective suitable for a business.
- "To make profit from the production and sale of laser diodes" is a commercial objective.

Objectives must be realistic:

- "To dominate the world market for laser diodes" is a commercial objective, but probably not a realistic one.
- "To achieve a and maintain over the long run a position of leadership in the world market for laser diodes" may well be a realistic commercial objective.

Objectives must be clearly defined in terms of measurable results:

- "To make money from the production and sale of laser diodes" or "To achieve a and maintain over the long run a position of leadership in the world market for laser diodes" are realistic commercial objectives, but they are not defined in terms of measurable results.
- "To make 22 % net return per year on capital employed from the production and sale of laser diodes" is a commercial objective that (we assume) is the realistic result of market studies, and the achievement of the objective can be measured.
- "To achieve a and maintain over the long run a 30 % share of the world market for laser diodes" is also a commercial objective that (we assume once more) is realistic and the achievement of which can be measured.

As a manager, I am more comfortable with the first definition – 22 % net return on capital – than with the second – 30 % market share. Why?

In the first case, all managers and employees of my company know that they must contribute to the 22 % return objective: by purchasing suitable inputs at the lowest price, producing in the most efficient manner laser diodes that can be sold at a target margin in a competitive market (i.e., they must be of a quality sufficient to allow setting a price that generates the target margin), marketing and selling the diodes at that margin, and replacing free of charge those diodes that do not achieve the guaranteed life time.

In the second case (30 % market share), the feed-back from marketing to production may well be: we must improve our laser diodes to achieve the target market share; the feed-back from marketing to sales may be: we must lower prices; marketing may also urge the service department to be more generous in accepting warranty claims so as to enhance buyer satisfaction. In short, striving for market share is likely to go at the expense of profit. How long will the company be able to afford a 30 % market share under these circumstances?

To keep the marketing men in check, the objective may be formulated differently: "To achieve the highest possible market share commensurate with a 22 % net annual return on capital".

I said that I was more comfortable with the objective "To make 22 % net return per year on capital employed from the production and sale of laser diodes", but I did not say that I was happy with it. If, as a member of the business club, the scientist-manager can completely forget that he is a scientist, he would perhaps formulate the objective as follows: "To make 22 % return per year on capital employed." That would leave it to the marketing people to tell him whether it is easier to make 22 % producing and selling laser diodes or other products, or shoes or golf balls, for that matter.

I have spent so much time on this subject, because setting realistic business objectives is one of the most important managerial decisions and because the manager who cannot forget that he is a scientist is likely to make mistakes under this heading that may seriously hurt his interests as a businessman.

3.2. BUSINESS PLANNING

Having defined his strategic objective (22 % return on capital), a manager needs to know how to plan its achievement. Under the centrally planned system this was, by definition, done centrally. In the market economy, each enterprise is responsible for its own business plan. Those that prepare a realistic plan are more likely to succeed in a competitive market place than those that do not plan or plan badly. And the top manager who knows what planning involves will be a better manager, because he will be able to tell a bad plan from a good one.

Starting from one or more clearly defined realistic strategic objectives, planning is a process of many stages some of which are repeated several times before the final plan is approved.

- Setting operational objectives.

  The strategic objective of earning a net annual return on capital of 22 % is commercial and it is clear: its achievement can be measured. However, it is not operational. Given only that objective, the purchasing department does not know whether it should buy the most expensive inputs to allow producing the finest laser diodes in the market, or the cheapest inputs to allow producing the least expensive diodes. Production does not know whether it is more likely to achieve the objective by producing fewer but better or more but inferior diodes. The service department will be inclined to keep service costs down – and risk offending customers.

  Setting operational objectives is an iterative process, i.e. some of its steps, or even the entire process will be repeated before the final decision is made. Top management defines the corporate or strategic objective (22 % return on capital), and each planning unit "translates" this into operational sub-objectives. For example, the purchasing manager will tell his department to plan for a 6 % reduction in the cost of inputs; production will be asked to cut down energy consumption by 15 % and reduce faulty products by 25 %; packaging will have to reduce costs by 8 %; the marketing manager will call for a reduction in the cost of advertising by 10 %; the sales department will be asked to improve margins by 6 % increase sales by 15 %; the service department will be instructed to reduce costs by 8 %; the personnel department will have to reduce total payroll costs by 12 %; and the finance department, finally, is told to plan a reduction of annual finance charges by 2 %.

This is where the matter stands after the first round of objective setting. The next steps in the planning process will show if the operational objectives are realistic.

- Identifying constraints.

  Constraints are obstacles to the achievement of the operational objectives. An objective that cannot be achieved is not realistic.

  In our example, the purchasing manager, who has to achieve a 6 % reduction in the cost of inputs knows that prices are going to increase through inflation by 4 %; in other words, the total reduction he has to achieve is 10 %. He knows that this is only possible through a combination of buying some inferior inputs, long-term supply contracts, pressure on some weaker suppliers and bulk purchases for cash payment from stronger ones. His departmental plan reflects this. He also knows that the weakest suppliers will not survive if he applies too much pressure.

  That causes problems for the production and the finance departments: Production is unhappy about the inferior quality of some inputs – remember, it was asked to reduce defective products by 25 % – and the change-over to long-term supply contracts, because planned changes in production technology will call for different inputs in six months time; the finance department – which has been asked to reduce financial charges – protests against the planned bulk purchases for cash, because they will call for higher working capital and, therefore, finance charges.

  Similarly, the sales manager says he can increase margins and sales only if the guaranteed life of each laser diode is extended from 6 to 24 months, because that is what the competition is doing, and if major customers are given six months credit at a low interest rate.

  You can guess what happens next: the service and finance departments – both under instruction to reduce costs – protest strongly. Service reports that an increase in the guaranteed life from 6 to 24 months will lead to a 200 % increase in the department's costs, unless the product is dramatically improved. Production vetoes this, unless it gets better quality inputs (this is vetoed by purchasing) and more manpower for quality control. And the finance department is adamant that six months credit to major customers is out of the question.

And so it goes. Each department examines the constraints that make the achievement of its operational sub-objectives difficult or impossible. The planning units carry out this examination against an examination of each unit's strengths and weaknesses.

- Identifying strengths and weaknesses.

  Companies tend to have competitive advantages ("strengths" in the market place) and competitive disadvantages ("weaknesses"). A manager who is ignorant of these may go through the motions of planning, but the result is not a business plan but a statement of hopes and wishes.

  Both the strategic objective and the operational objectives must be reviewed in the light of the company's strategic and operational strengths and weaknesses. A few examples:

  - If my product is weaker than those of the competition, how realistic is a strategic objective of a 22 % return on capital?
  - If my firm's relations with banks are weak because of bad past performance, and if the owners do not plan to make more capital available, how realistic is any operational objective that involves an increase in working capital?
  - If a competitor has a more modern production line and is using its production cost advantage in a price war, how realistic is an operational objective of an increase in sales without major price reductions (which in turn would threaten the achievement of the strategic objective)?
  - If I have the strongest service organization of all suppliers of laser diodes, why don't I increase the guaranteed life to 5,000 hours? Higher warranty costs will be compensated by additional market share won from competitors that cannot match the increase in the guaranteed life.
  - If I have the strongest sales organization, why not temporarily lower the strategic objective to 18 % return on capital and reduce prices until I have driven competitors out of the market? In two years I may be able to increase the strategic objective to 30 % return on capital.

  One great temptation for the scientist-manager who forgets that he is a manager first and scientist second is the failure to recognize the firm's relevant strengths and weaknesses.

  - It is no good planning a 22 % return on capital with a custom-made "Rolls Royce"-quality laser diode, if the market prefers to buy mass-produced diodes at "Volkswagen" or "Lada" prices.
  - It is no good letting pride in the product dictate the guaranteed life, if the resulting warranty costs will turn profits into losses.
  - It is no good planning a technically possible 15 % reduction in energy consumption when I know that I cannot get the financing for the investment necessary to achieve that saving.

- Identifying the options.

  Many business objectives, strategic as well as operational, can be achieved in more than one way. The alternatives are the business options. Options are derived from objectives. Sometimes the strategic options are as simple as "high volume at low price" or "high price but low volume" (Rolls Royce vs. Volkswagen); at other times they may be "domestic market only" or "domestic market and exports"; or "one product" vs. "diversification"; "internal growth" vs. "growth by acquisition", etc.

  The operational options are alternative ways of achieving operational objectives. Do I achieve energy savings by investing in new technology or by using a cheaper fuel? (This option may not be available when environmental legislation is a constraint.) Do I achieve payroll cost reductions by reducing the number of employees or by imposing salary cuts (employment laws may act as a constraint)? Do I increase manufacturing productivity by changing to a bonus system or by buying new equipment, or by doing both? Do I reduce the cost of packaging by subcontracting the manufacture of wooden cases now manufactured by the company, or do I change from wooden to plastic cases? Most planning units in a company will have several operational options.

- Costing the options.

  Having identified the different options, the planner will usually, but not always, select the one with the better "cost/benefit" ratio. Each option, even the option of doing nothing, has a cost associated with it and will generate measurable benefits.

  The planner has a problem when the option with the better long-term cost/benefit ratio is less attractive in the short term than another option. Do I invest now to improve the quality of my company's products and reputation and gain a long-term advantage over the competition but make losses in the next two years, or do I continue to produce low-quality goods as long as the market does not ask for better ones? Do I train my employees at the company's expense and risk that they are hired away by competitors but have a better and more loyal work force, or do I wait until better trained workers are produced by the education system? Do I go for maximum automation of the production process and risk dissatisfaction of unemployed workers and in the community but have a better product, or do I keep to present labor-intensive methods and risk that the competition gains an advantage?

- Consolidating the plan.

  Some, possibly all of these steps will have to be repeated until a realistic and internally consistent plan evolves and can be presented for adoption.

  A realistic plan means that every selected option is capable of being implemented, i.e. that the company has the resources (people and money) to do what the option calls for, and that there are no other barriers to its realisation. A plan is internally consistent if the implementation of an option in any one department does not prevent the implementation of an option in another. If the company has sufficient financing only for an increase in working capital or for an energy-saving investment, it is inconsistent (and unrealistic) to plan for both.

Management training must also deal with plan implementation and performance monitoring, i.e., the activities following the formal adoption of the plan.

3.3. PLAN IMPLEMENTATION

Plan implementation is a matter of both organization and communication.

- Without an organization designed for plan implementation planning remains a wasted effort. Key to such an organization are allocation of responsibility and delegation of authority. Both are inseparable elements of planning.
    - Responsibility for every aspect of the plan must be clearly defined. It will not do for the top manager to assume responsibility for plan implementation and the achievement of the operational objectives, because he or she will not purchase materials and produce, package, market, sell, distribute and service the goods. The managers responsible for those activities must be responsible for achieving plan targets.
    - Those responsible for the achievement of the plan must have real authority to take the necessary action. Some Russian managers find this hard to accept, because the management culture of the past centralized all authority in the top manager.
- Communication in the context of planning refers to the systems and procedures required to collect operational data and process them into information, and distribute the information in a timely manner to all those with responsibility for plan implementation. We are talking of the hard- and software and the processes that constitute the management information system.
    - It is essential, once again, that information does not flow one-directionally from the bottom of the company hierarchy to the top, but also in the opposite direction, as well as laterally, i.e., between departments and divisions. Every manager must have the information allowing him to meet his responsibilities and exercise his authority purposefully.

3.4. PERFORMANCE MONITORING

Planning without performance monitoring is a waste of time and undermines the credibility of management.

- It is a waste of time because nobody will know whether or not the plan targets are being achieved and, if not, why not.
- It undermines the credibility of management, because the people who created an organization for planning, and inflicted the pressure of the process on all those engaged in it, now appear to disavow the effort. The next time this management will call for an extra effort, the employees are bound to react with less enthusiasm.

A good Management Information System ("MIS") identifies company-specific critical indicators to be monitored by individual departments. For example, the Sales Department may monitor total sales, market share and net revenue per unit sold; the Finance Department may monitor liquidity; the Accounts Department accounts receivable, etc.

- Those monitoring the plan will compare actual result with plan targets. An "acceptable deviation" should be defined for each critical indicator, and the monitors will treat any result that falls within the acceptable range as "routine" and devote their time and attention to "unacceptable" deviations, including those that are positive beyond expectations. A series of surprisingly positive results may call for an early upward revision of plan targets.

3.5. EFFECTIVENESS TRAINING

Remember our definition of managing as Planning, Implementing and Controlling the Effective and Efficient Use of Resources in the Pursuit of Clearly Defined, Realistic Objectives.

Effective action is goal-oriented action that is successful. A brief word on both these requirements

- Management training must include effectiveness training. The action of managers must be determined by goals and objectives. Effectiveness training focuses on defining realistic objectives and on acting purposefully. Both are essential skills for managers, including scientists as managers.

    - The scientist as manager who continues to be guided by his pursuit of a Nobel Prize may be an effective scientist but is probably not an effective manger – unless the award of the Nobel Prize is also a genuine commercial business objective, for example for a consulting firm that markets the "brains" of its employees.

        - Another example: the commercial film studio in Hollywood that seeks to maximize shareholder value must produce "Arnold Schwarzenegger" films, not films of high artistic value. The studio boss who seeks artistic excellence may be respected as a cultured man, but pitied as a businessman.

        - But the action must also be successful to be effective. The business world is littered with unsuccessful attempts at effective action (the Edsel and the Pinto, the film "Water World"). Unsuccessful but purposeful action is, in the end, of no more use than action not guided by purpose.

3.6. EFFICIENCY TRAINING

- People and money are the basic business resources.
- Efficient action maximizes the productivity of the resources employed in the pursuit of a business objective, thereby minimizing the consumption of resources. Efficient resource management can actually increase the resources at the company's disposal.
    - Efficient action can increase the resource "people" without hiring personnel:
        - training will add to their skills;
        - responsibility will add to their experience;
        - incentives will add to their motivation;
        - safety at work will protect or increase their health.
    - Efficient action can increase the productivity of the resource "money":
        - for example, by putting the right premium on "money today" (as against "money later"), the efficient manager increases the productivity of the resource.

4. Conclusion

My three messages to the scientist-managers among you were:

(I) Appreciate that the "business club" values your management skills more than your achievements as a scientist.

(II) Therefore, learn the rules of the business club.

(III) And acquire the skills that allow you to play – and win – by the rules of that club.

If you have already done all this, you will have found nothing new in my speech. If you have not, I hope I have made you think.

PECULIARITIES OF INNOVATIVE ACTIVITIES OF SMALL AND MEDIUM HIGH-TECH ENTERPRISES

Andrei FURSENKO, Vladimir SPIVAK
SPb. Regional Foundation for Science & Technological Development

Successful operation and development of small and medium enterprises (SMEs) proves to be an important indicator of economy health and social stability of the society. This is connected with the fact that SMEs are inseparable from the occurrence of the middle class not just deeply interested in the progressive, evolutionary development of the society, but also capable of ensuring this development within the framework of democratic approaches.

SMEs' development is especially important for high-tech sphere where the greatest material and intellectual resources of the former Soviet time were concentrated. About 5 years ago highly qualified specialists began to leave Russian high-tech industry, first of all, the defence industry. Some of them emigrated, others – completely changed the profile of their activities. However, the significant number of them went to newly established SMEs and continued the work with their R&Ds and their introduction to production.

By the data of the State Statistics Committee in the end of 1995 there were about 900 thousand small enterprises where worked about 9 million people or 15.7 % of the total number of people employed at different enterprises. Under secondary employment conditions the same year small enterprises used the labor of more than 5 million people. In St. Petersburg small high-tech enterprises made about 10 % of all (nonindustrial ones included) SMEs.

Lack of information on high-tech SMEs and their activities, insignificant age (sometimes less than 3 years) and absence of market skills suffered by their managers are added to general present-day problems of attracting investments in this sphere. The situation is aggravated by the fact that investments – the key element of any developing economy – are connected with a greater risk in high-tech sphere and have a long payback time. Though to be fair one should emphasize that in case of success profit here may be significantly higher than in other industries.

Despite the latter, and the widely declared support of small businesses, federal and regional authorities provide no funding for complex high-tech production. At the same time private investors (both Russian and foreign) refuse to invest in high-tech SMEs not only because of the long payback time, but also because it is difficult to assess the economic effectiveness of high-tech projects.

The development of innovation activities (IA) and the preservation of the high-tech potential make state priorities in applied research for all developed countries, but in

C.M.R. Verkoeyen et al. (eds.), Management Training in High-Tech and R&D, 51–55.

Russia today they face a great number of problems the major part of which are of objective, global character and have to be solved at the federal level.

Among them one should mention taxation policy hostile to manufacturers, exhaustion of the Soviet technological base and lack of resources for its replenishment, weakness of the Law in intellectual property, limited financial sources.

As far as subjective difficulties are concerned, i.e. those which may be settled by the participants of R&D and production activities themselves, provided they get certain support of federal and regional authorities and international community, they include the insufficient development of infrastructure for IA support and absence of appropriate management at all its stages.

The main reason why unique R&Ds and scientific ideas remain in no demand consists in the fact that the old high-tech infrastructure (Academy of Sciences, Ministries and their research institutes, universities) has turned out to be absolutely unfit to market economy challenges, and a new one is just being established.

The St. Petersburg Regional Foundation for Scientific and Technological Development (RF STD) established in 1992 as a non-profit organization was one of the elements of the new regional infrastructure for IA support. Since the very beginning of its work the support and development of IA has been the point, i.e. those applied research which can be used in production, commercialized and become a part of standard market economy.

The first task we faced in this work was the formation of selection principles and criteria for the support of innovation projects. Initially these principles were formulated as follows: funding is provided for regionally oriented high-tech projects having the implementation term not longer than 2-3 years, the funding is given mainly on credit basis.

Within the process of our work the principles were adjusted and developed, but the main approach remained unchanged: funding should be provided under preferential terms and on credit basis.

During the last four years RF STD got more than 400 applications of which:

22 % from large enterprises and organizations (academic and branch institutes, defence design bureaus, higher educational institutions);

30 % from small enterprises dependent on large enterprises (actually these small enterprises are not very much of independent structures, they are rather independent banking accounts opened by large enterprises to solve their economic problems, like avoiding taxes, redistribution of wages, etc.);

2 % from individuals and, finally,

46 % from really independent small enterprises established mainly by scientists and engineers for their own R&Ds promotion.

By our estimation the latter make not more than 10% of all high-tech SMEs. Finding themselves at the initial stage in the most complicated situation they demonstrate maximum efficiency and tenacity vividly showing all IA opportunities and problems.

Due to our assistance projects got funding from two main sources: the Russian Foundation for Technological Development of the Ministry of Science and Technological

Policy of the Russian Federation and the Foundation for Assistance in Small High-Tech Businesses Development. 48 projects have already got funding. 55 % of them are projects of independent small enterprises, 13 % – were introduced by large enterprises, and 32 % – by small enterprises dependent on large ones. As far as individuals are concerned, those of them whose projects were of real interest, have established independent companies within the process of their projects development and now they have got funding.

Among specific IA problems of independent high-tech SMEs one should mention the following:

- high prices and instability of rent terms;
- official and unofficial costs for ensuring the security of both property and enterprises themselves;
- needs in costly technological and measuring equipment which they cannot or do not want to purchase as their full property;
- strong dependence on partners and suppliers, as small enterprises need cooperation even more than large ones;
- absence of guarantees for commercial credits and investments required, on one hand, and practically absence of possibility to obtain budget funding, on the other hand;
- insufficiency of managers' experience for managing enterprises under market conditions, etc.

During the first three years of its operation RF STD focused its attention on the search, selection and financing of innovation projects, as well as on project managers training and consulting (for which purpose a specialized Center of High-Tech Industry Management Training was established).

However, the acute problem of high-tech SMEs location in equipped premises for sufficiently long time (3-5 years) under steady conditions and at reasonable rent has not been solved.

This is not purely Russian problem. All over the world, independently of economic development and political system, special research parks, business-innovation centers, incubators are established for the support of IA and high-tech SMEs. As a rule they are connected with different universities, but they may exist independently as well. Until now the attempts to establish such structures in Russia were anything, but success: at best, offices of various high-tech firms were accumulated in a single building. The main mistake made by the founders of our research parks was that they began their activities either with getting real estate (very often unsuitable for production and technological activities), or with setting-up the formal frame structure of directorate with the hope of getting large investments for building specialized premises, as it is done in economically developed countries.

It is evident that the main requirement to any project is the adequacy of its scale and opportunities available. Resources in Russia being limited, one should concentrate not on building new, but on the adaptation of that available (premises, equipment, specialists) to the conditions of the present day. In this process one should proceed exclusively from market needs, i.e. demands of certain enterprises for production premises, equipment, services.

From our point of view the solution may be found within the framework of conversion at large defence enterprises where large premises, expensive technological equipment and highly qualified personnel get free.

We tried to solve these problems within the framework of the Agreement on providing RF STD with additional functions of the Innovation Technological Center (ITC) which was signed about a year ago by the Minister of Science and Technological Policy of the RF B.G. Saltykov and the Mayor of St. Petersburg A.A. Sobchak. Pursuant to this agreement in 1995 one of the largest enterprises of the city – "Svetlana, PLC" which is under conversion now gave RF STD, as a voluntary property contribution, one of its technological buildings. Due to this step "Svetlana", on one hand, managed to get rid of taxes on this property and costs for its maintenance and, on the other hand, may expect fruitful cooperation between its production and high-tech SME's located in ITC.

The building of 7,000 squ.m total area needed some replanning and reconstruction in order to ensure the rent of about 5,000 squ.m. And in less than a year due to the support of the Ministry of Science and Technological Policy of the Russian Federation, the Foundation for Assistance in Small High-Tech Businesses Development which provided RF STD with a preferential credit and due to the funds provided by high-tech SME's – future tenants the building has been practically brought to the necessary working condition. It is very important that these are not offices which are located there, but complex pilot technological productions of SME's in opto- and micro-electronics, new materials, machine-building, etc.

It is for the first time in the city, and may be in Russia as a whole, that we managed to locate a big group of SMEs on a single territory, moreover that they are united not by corporative interests and origin, but by common motivation to develop high-tech products. Their diversity may provide a great synergistic effect. By the end of 1996 the number of SMEs located in ITC on the basis of their business-plans competition will make about 15, in total their number may amount to 20. It is interesting that ITC characteristics given here correspond to average statistic characteristics of similar institutions in GFR.

The main task we have to solve in the nearest future is acquiring power-supply independence of "Svetlana"'s infrastructure, as now this company is not having its best time. But this task does not look insoluble to us either, as we feel real support from both city and federal authorities and from "Svetlana".

The purpose of RF STD in establishing ITC was to meet the main requirement of high-tech SME to infrastructure – complex approach and high quality of services rendered. Along with providing SMEs with specialized production premises equipped with all necessary technological systems the following services are implied for the tenants:

- search and ensuring funding for R&Ds conducted by SMEs and their monitoring;
- informational, telecommunicational and consulting services;
- management training;
- establishing cooperative equipment stock for research and R&D's;
- intermediary services in contacts with international research organizations and setting partner relationship with foreign companies.

To provide the full picture of present-day high-tech SME's one should mention that lately there have occurred serious tendencies towards changes in their needs.

Fixed assets. Five years ago they had no ideas, but one, that they should use the premises available in mother organizations free of charge. Today independent SME's are more interested in legal relations: either they get premises as their property, or lease them. Then they feel themselves more confident, it is easier for them to negotiate with partners on joint activities – no serious investor will get interested in a company which is located in other company's premises having no rights to them.

Finance. In 1991–92 all organizations saw the only opportunity for themselves in getting budget funding, at least a grant for a project, for the use of which they had to report. Today high-tech SMEs ask for credits, though insist of privileges. They are also ready to participate in serious partner relations, ready to make investors their shareholders, to accept contributions to their Registered capital, realizing that there are no other ways.

Services. About 4–5 years ago the magic word "business-plan" was the utmost dream. Business-plans were drawn up without full understanding of what it meant. Today SME's make absolutely concrete inquiries for informational services, market research and deep financial analysis.

Partnership. Earlier in their search for partners SME's hoped to find a "money bag". Today the most advanced enterprises are looking for technical partners or for those who can participate in production together with them, or in the sales of their products.

Summarizing everything said above we can make a conclusion that in spite of all difficulties there exist quite serious positive changes in IA development in Russia, first and foremost, in understanding of the situation by the subjects of these activities. And the most promising subjects prove to be independent high-tech SME's which can make a basis for the new high-tech industry of the country (not so large-scale, as in the former USSR, but more dynamic and fully adapted to market economy), provided they get certain support of the state.

SOME PROBLEMS OF INVESTMENTS IN RUSSIAN HIGH-TECH PRODUCTION

Valentin LEVITSKI
The Russian Technology Fund L.P., St. Petersburg, Russia

The Russian Technology Fund (RTF) is an international fund of long-term risk (venture) investments in small and medium firms dealing in manufacturing competitive high-tech products chiefly intended for the Russian market.

Legally RTF is a limited partnership registered in the USA in 1995.

In its operation RTF is mainly confined to the economic region of St. Petersburg.

The start-up capital of RTF makes more than 10 million USD. Its capital growth is restricted by the amount of 30 million USD.

The founders of RTF are :

- "Top Technology Limited" (TTL), a British venture-capital company having the experience of investments in more than 120 British and American firms;
- "SITRA", National Research and Development Fund of Finland.

Among the investors of the fund one should also mention:

- European Bank of Reconstruction and Development (EBRD);
- International Finance Corporation;
- Rothschild's company ("NM Rothschild and Sons Ltd."), Great Britain;
- a number of Western-European and American private investors and firms;
- Russian Trade-Industrial Bank, St. Petersburg.

In St. Petersburg RTF is looking for opportunities of mutually beneficial cooperation with small and medium firms which:

- find themselves at the stage of commercial production and its expansion;
- have strong position on the market and well-grounded growth prospects;
- introduce new technologies.

The essence of venture investments consists in the fact that an investor invests his capital in the registered capital of a firm and thus becomes one of its owners – a shareholder. In this process RTF is not striving to get the control block of shares and, being a share-partner in business and having interest in the firm's success, it most directly takes part in the management and development of the firm it invests in.

The commercial interest of RTF lies in the successful development of the firm, so that in 3–7 years its value (i.e. the value of its shares) could be increased by 6–10 times, which will make in average 40 % of annual growth. That allows RTF to sell, at a certain

C.M.R. Verkoeyen et al. (eds.), Management Training in High-Tech and R&D, 57–59.

stage, its shares either to the firm itself, or on the Stock Exchange, or to a third party interested. As a result, RTF will not just get back its investments, but will also gain profit on the capital.

The success in financial activities of a venture capitalist is not guaranteed – in case of a failure he suffers risk like any other shareholder. In case of success the main result will be not only profit got by RTF, but also the company's prosperity.

In choosing a portfolio company RTF proceeds, first and foremost, from the company's management level and the ability to develop a clear business-plan oriented to the final result.

RTF does not bear technical risk, ***it invests not in ideas, but in people***.

Let me dwell upon this last statement. The practice of work with firms provides quite a typical for Russia picture of quality and quantity development of small high-tech manufacturing business. First, the core of all firms is usually constituted by outstanding engineers and scientists having accumulated significant experience of organizational work in Soviet times who try to implement their R&D's outside the bureaucratic restrictions of governmental institutions. Second, the number of small firms established with direct participation of "mother-organizations", i.e. large governmental or joint-stock manufacturing or research enterprises mainly of defence orientation, is increasing. Such small firms are known as "spin-offs" or "splinter" firms. Having an opportunity to use the infrastructure and scientific potential of their mother-organization they get quite favorable conditions, especially for the start-up stage of their activities.

The third positive trend proves to be the change in the mentality of scientists and researchers who are no longer striving to get "fast" money by selling their "brains" or R&D's "to the West", but see their aim in establishing state-of-the-art production in Russia on the basis of the scientific potential available, in competing with not just domestic manufacturers, but with foreign ones as well, not just on the domestic market, but on the international one as well.

In this connection many of the businessmen having started their own business become, in fact, entrepreneurs combining at least three features in themselves: ability to creative activities, reasonable risk and skillful management, that in finance as well.

Venture capital is a kind of an entrepreneur in the field of finance. Both entrepreneurs have one and the same final purpose: to create wealth either by manufacturing, or by financing activities. To achieve this purpose they badly need each other. From here stem both fruitfulness and effectiveness of their cooperation proved by the long development practice of small high-tech business and venture capital in the USA and Western Europe. Therefore, in choosing investment object RTF primarily orients to the availability of entrepreneur skills in the firms' management teams, i.e. after all, it invests in entrepreneurs, not in R&D's.

Being implemented in Russia such an approach, though typical of venture capitalists in the West, is hindered by a number of specific problems characteristic of the today's stage in restructuring domestic economy and, correspondingly, of entrepreneur training.

Among these problems one should mention the following:

1. Hostile attitude of a firm's management team to the very idea of conceding (selling) some part of shares in the firm to RTF, of RTF's participation in the firm management, in the work of the Board of Directors. By our opinion, this is caused not

by any peculiarities in the national culture of Russian businessmen (like it is, for instance, in Latin America or Germany), but by mere absence of this culture, which can be explained historically and, as practice shows, is quite temporary.

2. Significant overestimation of a firm's "value" by its management team, of the importance of their innovation and intellectual property, which makes drawing-up of a mutually beneficial deal scheme more difficult. To do justice to Russian small high-tech firms, it should be said that this is typical not only of them.

3. Failure to understand the importance of business strategy development and, consequently, failure to understand the role attached to business-planning, most often considered to be a luxury, not a necessity. As a result firms managers cannot set realistic objectives, do not understand the importance of actual risks estimation, do not pay appropriate attention and efforts to market research, i.e. to the analysis of its volume, development trends, competition, competitors' strengths and weaknesses, to pricing policy development, etc.

4. Lack of knowledge in finance management. In practical work it is often diminished to simple accounting, without any financial simulation or financial analysis of the firm's operation.

Some of the main problems touched upon in the presentation which are faced by RTF in its investment activities are of quite evident and objective character, though sometimes aggravated by a number of subjective factors, such as the origin of entrepreneurs from scientific and research community. Therefore, the task of management training and re-training, both regular and on-the-job, appears to be quite urgent.

Becoming a share-partner in the business of a portfolio firm RTF is striving to bring added value to the company via participation in hands-on-management and on-the-job training. First and foremost, that means assistance in drawing-up realistic business-plan, in financial analysis of the firm's activities, in financial simulation of its development after getting investments, in market research, in introducing integrated management program in order to timely meet ever changing challenges of the business environment.

Besides, RTF collaborates with a number of Russian and international institutions and programs making contribution to the improvement of business-training for Russian entrepreneurs many of whom take part in the work of the present Workshop.

The topic of the Workshop and the makeup of its participants provide for an optimistic and simultaneously realistic forecast as follows: due to undertaking common efforts aimed at improving Russian entrepreneurship culture the majority of the above mentioned problems faced by investors will be eliminated in the nearest future and, consequently, investments will increase.

THE DIFFICULT, BUT REWARDING WAY OF TURNING R&D RESULTS INTO BUSINESS SUCCESS

Manfred F. GARDIAN
Deutsche Management Akademie, Niedersachsen

Distinguished ladies and gentlemen,

I would first of all like to express my heartfelt thanks to the event organizers for the invitation; it gives me the opportunity to talk, within the framework of the given topic, about the concept of the German Management Academy of Lower Saxony Non-Profit Company (DMAN) Economic Development through Management Training.

I promise you that I won't indulge in a long academic exposition, but will show you a way based on real-life experiences, and there may well be others, in which we can jointly train managers and trainers, who are able to convince by virtue of their experiences, with the simultaneous production of usable results.

- State of Lower Saxony
- Volkswagen AG
- Preussag AG
- Hannover & Hildesheim Chamber of Commerce and Industry
- Lüneburg-Wolfsburg Chamber of Commerce and Industry
- Norddeutsche Landesbank
- Deutsche Messe AG Hannover
- Lower Saxony Industrial Training Institution

Fig. 1. Shareholders of DMAN

C.M.R. Verkoeyen et al. (eds.), Management Training in High-Tech and R&D, 61–78.

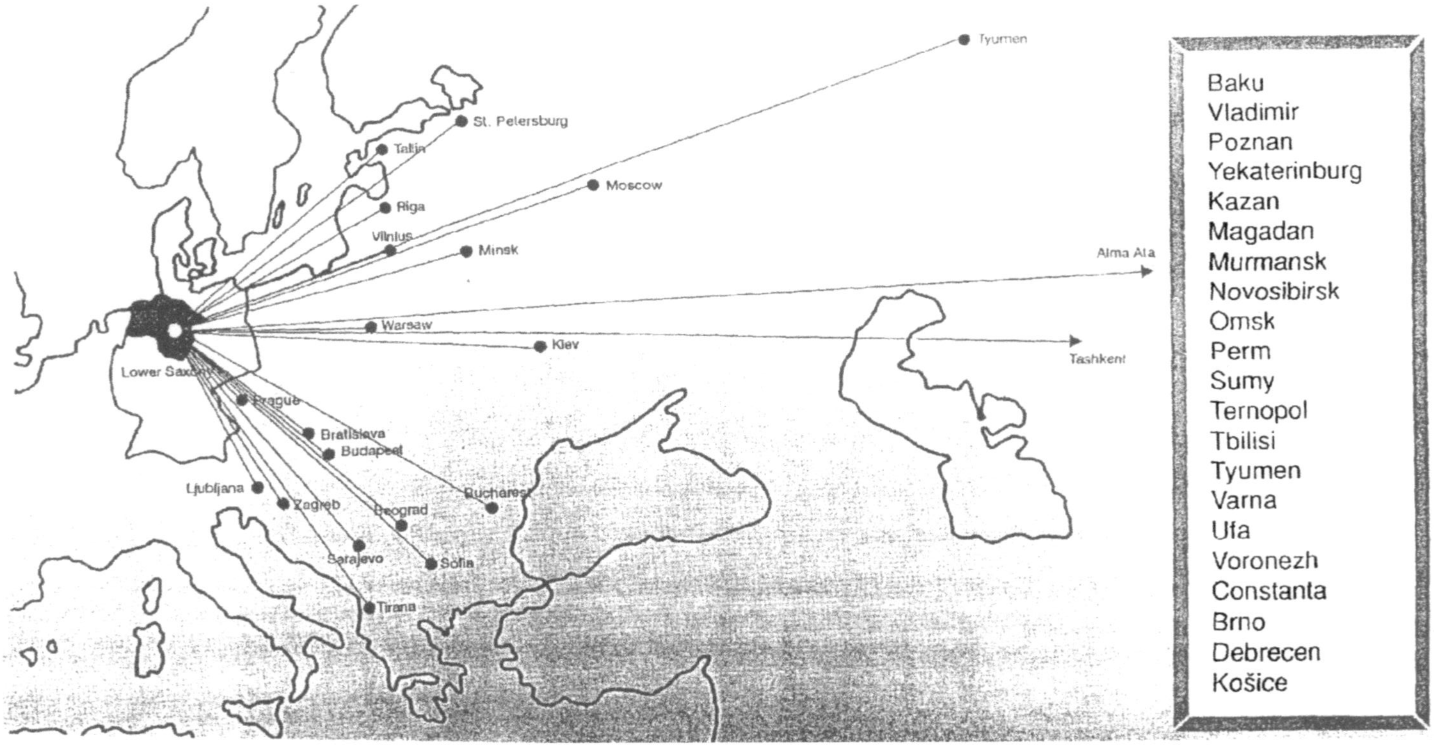

Fig. 2. Partners of DMAN

First a few words about the institution which I represent and about me personally.

The corporate structure (Fig.1) of the non-profit Company [limited liability co.] German Management Academy of Lower Saxony shows its closeness to business. The two large, producing companies of Lower Saxony are represented, with Volkswagen AG and Preussag AG.

The Norddeutsche Landksbank and the Deutsche Messe AG represent the circle of service-providers which is becoming larger and larger. The Chambers of Industry and Commerce get the Academy the access to the large circle of small and medium-sized enterprises in Germany. I may recall, at this point, that approx. 80 % of the gross national product in Germany is produced by small and medium-sized enterprises. The closeness to business is also documented by the fact that the Academy director is a man of business and has "suffered through" management training in his professional development.

Over 7,500 executives have been trained in Celle since the founding of the Academy, who all come from the area of the former Comecon (Fig.2). It is perhaps interesting for you that in at least 6 cases we have already trained up to 200 executives and executives to be of a single Russian company.

Our target groups are the today's and future decision-makers or opinion-makers (Fig.3). The volume is distributed about half and half between these two groups; through the train-the-trainer program, we also contribute to the fact that training modules can take place in your country to an increasing degree.

- **Current decision-makers** — 40–60 years
  - **1. Top-tier management**
  - **Functional level**
    - marketing/sales
    - production
    - personnel
    - finances
- **Future decision-makers** — 25–40 years
  - **Next generation of managers for tiers 1 and 2**
- **Future management trainer**

Fig. 3. Order of Priority for Participants

The key to the high acceptance of our work in your country lies in the orientation to industrial sectors, among other things. Trainers who come from the German industrial sector corresponding to that of the participants are used by us in the project work or in the seminars, so that branch insiders teach branch insiders.

Let me briefly make clear, with an example, what we mean with *Economic development through management training.*

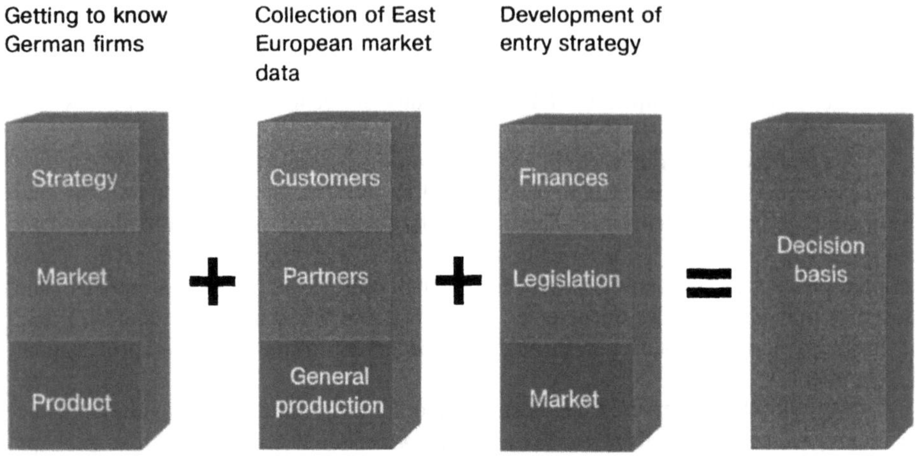

Fig. 4. Celle Foundation: Entry Strategy

Within the framework of the project "Celle Foundation" (Fig.4), a group of Russian executives first get to know a German enterprise: product, market, competitors, etc., in order to then subsequently work out with us which market data are important for this entrepreneur, who is faced with the question as to whether he should invest into your country. The task for the participants of this project is to then gather these data themselves in a region or even in several regions, and to structure them in a preliminary way, so that entry alternatives to your market place for this entrepreneur can be recognized. In a third step, the participants, together with us, will structure the gathered data in such a way that these desired entry alternatives become very transparent, and the participants will personally present the result of their work. In the process, the participants learn to develop a business strategy, in addition to how a entrepreneur thinks. At the end of this process, there is then a basis for decision, on which the entrepreneur can build with his next business steps. Thus, we connect the seminar/training work with the production of concrete results, which can induce a businessman to look for cooperation with enterprises in your country.

A second essential feature of your work is that we spend a great deal of time in detailed exchanges with the recipients of our service, in the run-up to the actual project, in order to insure that we structure the project contents in such a way that they meet the needs of the recipients (Fig.5).

This comprehensive groundwork (Fig.6) also becomes especially understandable when I tell you that the participant's own participation in our turn over, in foreign currency, has already reached over 50 %, and that is why we view the seminar participants as customers, who retain their right to constructive criticism through their own participation. One cannot achieve the conversion of a planned economy by opening up subsidy pots. One has to want change, even with the risks connected with it.

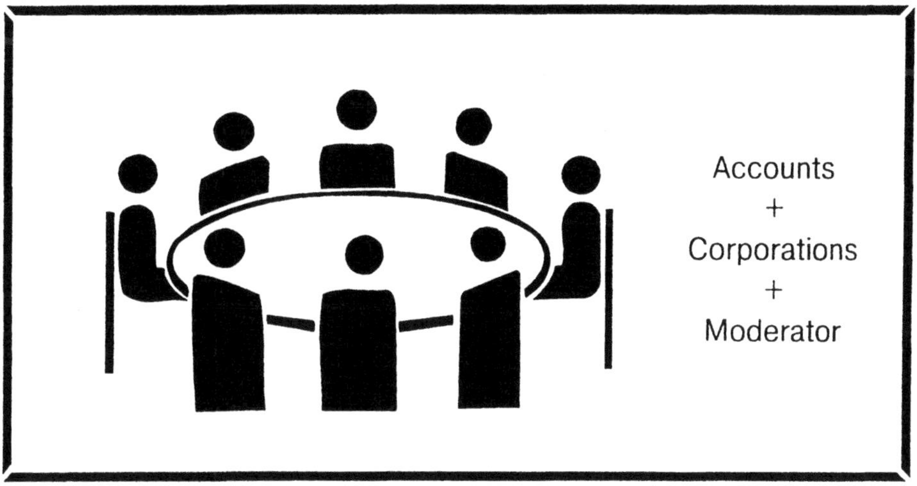

Fig. 5. Celle Castle Round: Account Nurturing/Solicitation

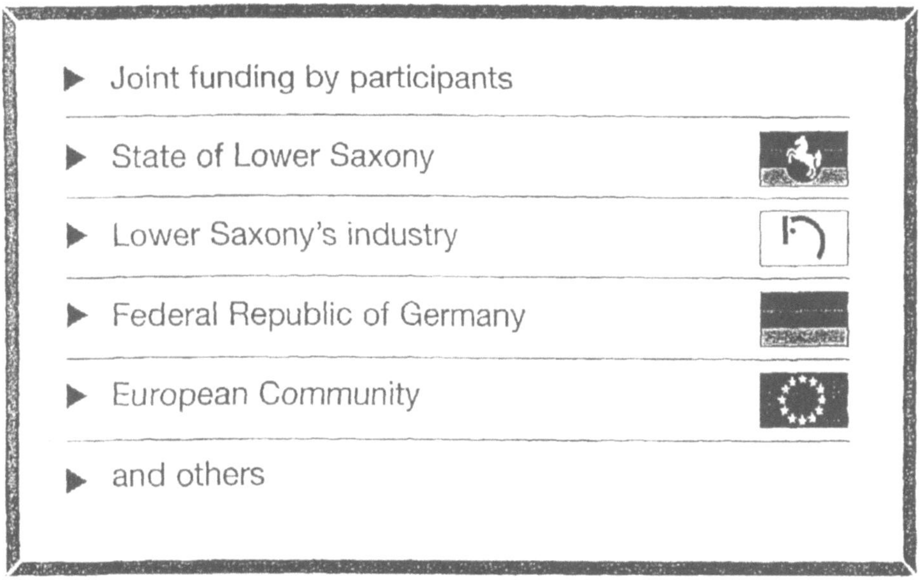

Fig. 6. Funding Principle: "Co-Financing"

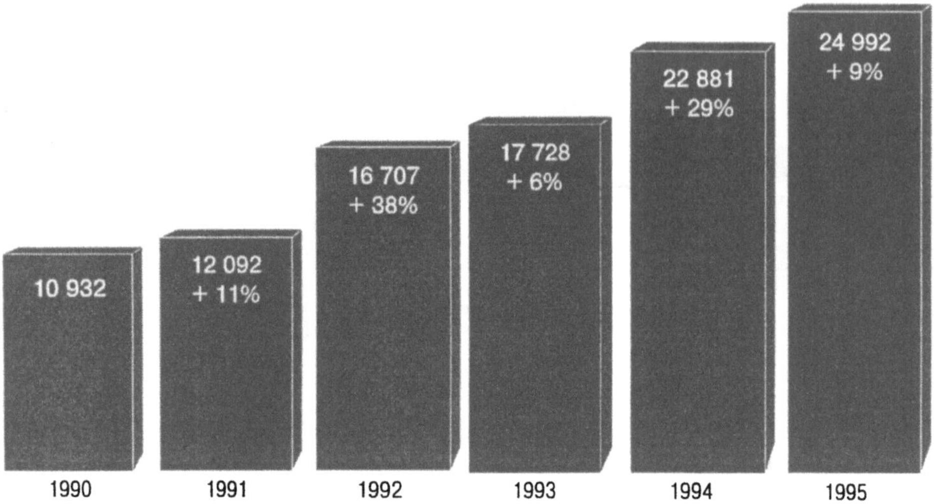

Fig. 7. Growth in Participation (Participant Days)

The fact that we have brought about such a rapid development of participants (Fig.7), despite increasing their own increasing cost share – we had 120 people daily in the Academy building in 1995 on an annual average – points out that we are meeting the expectations of the people that come to us more and more precisely. The ability to listen, in order to really understand the needs of the client, is one of the essential requirements for successful business activities (Fig.8).

▶ **Deutsche Management Akademie Niedersachsen**

provides services for

- ▶ Policy and Administration
- ▶ German economy
- ▶ Economies in Central and East European states

"We meet customers' requirements"

Fig. 8. Self-Conception

The image that we have of ourselves results from this. We offer the service of our abilities to our clients, from the circle of your enterprises, from the circle of German industrial undertakings, but also where requested of us for the tasks that arise in the area of politics. We do not belong to the circle of know-it-alls, because we are convinced that there will be a particular path, appropriate to your circumstances, into a socially-tolerable form of a market economy.

A few words about me personally, to the extent that facts from my professional development are important for the given topic:

When you look at my development, you find regularly-occurring change, with an ever-increasing responsibility, between production, technology, technology and marketing, distribution, production and technology, controlling and purchasing. In every case, responsibility for personnel was associated with these functions and tasks (Fig.9).

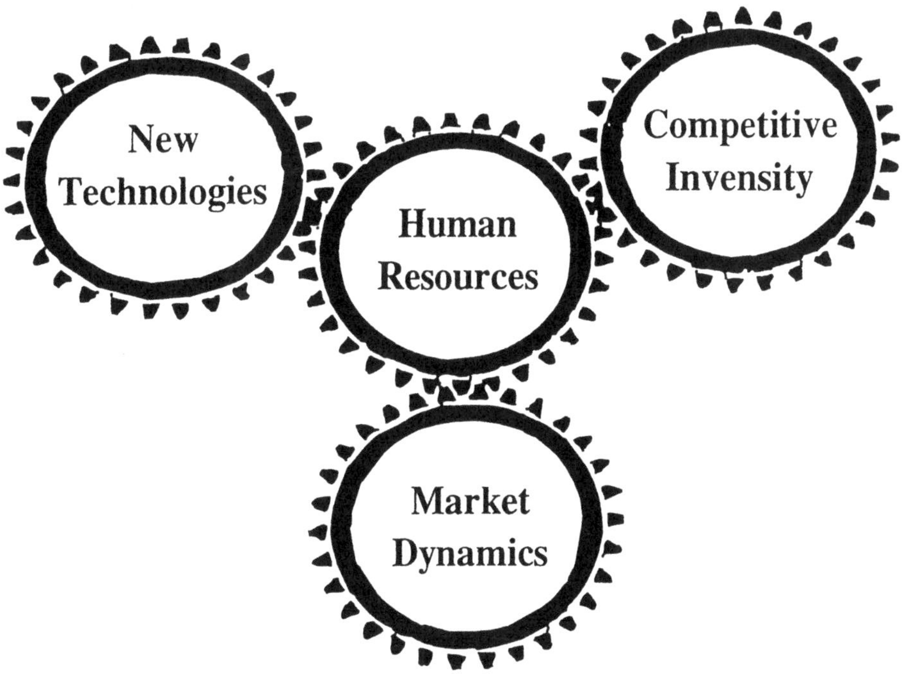

Fig. 9. Human Resources as The Driving Force for Innovations

Being able to listen, in order to understand the needs of the customer, and the Human-resource will play a role in my remarks again and again. You have a person in front of you who has acquired his knowledge through the path of "learning by doing". My personal training, e.g. at INSEAD, was essentially for securing the theoretical basis for my given tasks.

Somewhat self-ironic:

I am a generalist who knows something about everything, nothing completely about anything, and that is why I am convinced that only a team, in which abilities have assembled that complement each other, is successful with goal-directed management. Only the bundled forces of a team will be the impetus for innovation and avoid the risk that they are driven by competitive intensity, market dynamics and/or new technologies.

There is a second fundamental conviction; the alignment to a business goal is a requirement for successful continuing education and a successful RD&E activity. If this alignment does not exist, only an intellectual self-satisfaction and/or a waste of money will result from the education or development activity. There are few institutions in Eastern and Western Europe that can really afford that.

A third fundamental conviction influences my work: development activities have to be aligned to the actual or assumed needs of potential users. Development activity is always the basis for future business. The customers and not the enthusiasm for technology, are at the centre of attention of our actions.

The starting point of my reflections is that a group of developers are convinced that their development result can be put to a broad application by various users. Because they are convinced that their development result is unique, they also believe that one can earn a lot of money with it. It has to be made clear with these considerations, first of all, that the use of new technologies is always the start of a long route to market maturity, which requires a great deal of patience at the beginning (Fig.10). Many developers immediately set out in the search for a potential partner in their enthusiasm for their result.

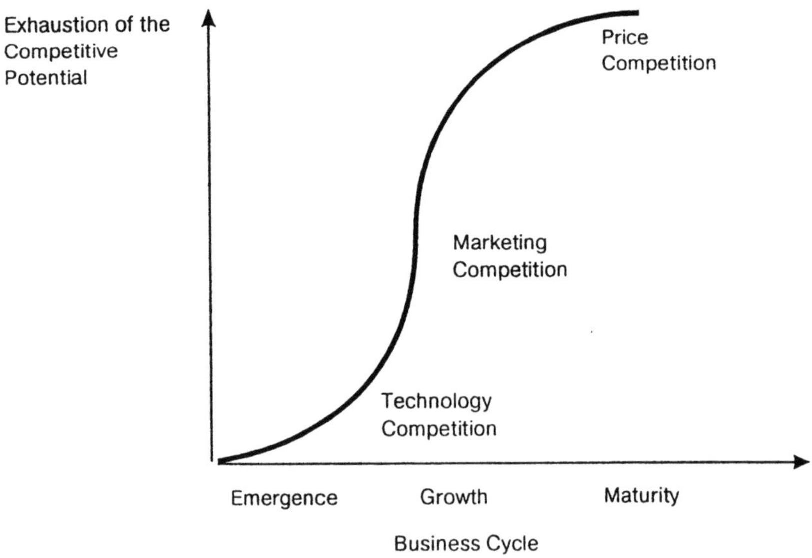

Fig. 10. Life-Cycle Phase and Market Growth

Point of view of researchers — Point of view of managers

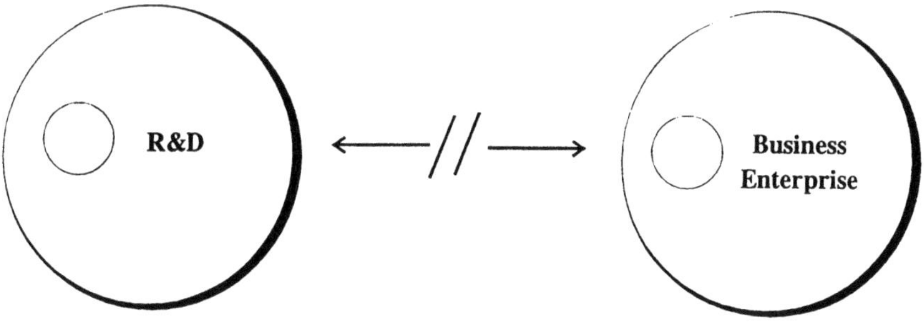

Culture of R&D-oriented areas

- Orientation to the world of science and the goals of technical progress
- R&D areas as a part of the world of science
- Individuality and independence of the researchers

Culture of market-oriented areas

- Orientation to the world of business and economic goals
- Business enterprise as a part of the world of business
- Integration of the managers in a hiperarchical organization structure

Fig. 11. Point of View of Researchers – Point of View of Managers

In so doing, they meet up, not with colleagues as a general rule, but with managers, who assess development results quite differently, with their way of looking at things, than the developers who are oriented more to the technical progress (Fig.11). This somewhat spontaneous behavior according to the rule "it's true we don't know where we want to go, but we're marching off anyway", is seen very frequently, but is fundamentally wrong.

Rule number 1 in this situation is protecting one's own know-how and the ability to analyse foreign patents, in order to be able to judge one's own strong points better. A process, by the way, which should accompany the development work from the very start. Knowledge of the legal situation is part of this and one has to be able to deal with the tools of the world-wide patent search and the patent assessment.

In this phase of a development project, the rule "talk is silver, silence is golden" applies.

I would not like to go into patent legislation and the handling of this in Russia at this point, but it is a field that has to be taken into consideration with the training measures. I mean with that not so much the formal legal side as daily handling and the consequences that result from this daily handling. A trainer workshop for this specialised area with trainer from Russia and from the West appears to me to be a suitable instrument to develop a sound training block.

Now after the R&D-people have done this homework, they are as nervous as racehorses as a general rule and would like to go out into the wide world again, in order to market their invention, aware of their strong points. Would you let your people loose now? Of course not!

One of the most difficult questions is yet to be answered: where are the greatest commercial success potentials in actuality? (Fig.12).

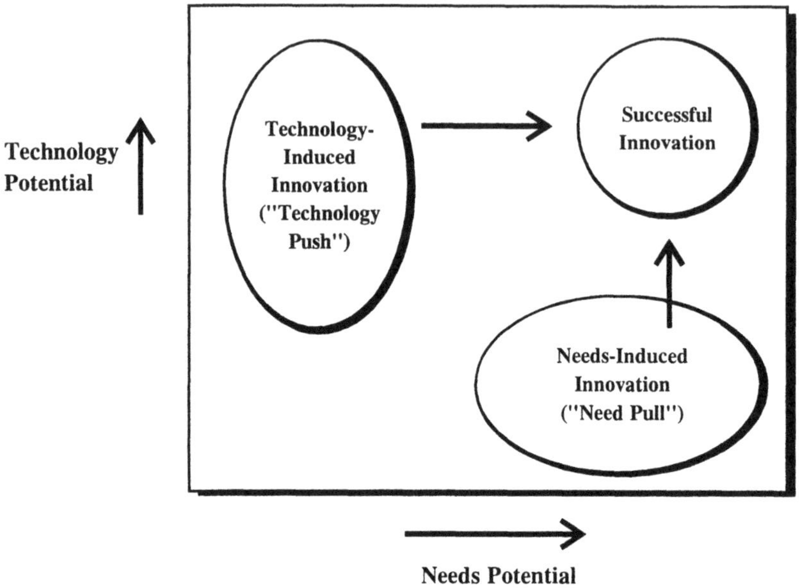

Fig. 12. Strategic Technology Management: Coordination of Technology and Needs Potentials

Creativity, vision and imagination are called for, in order to first identify possible areas of a successful commercial implementation.

Creativity and vision only reveal their full power when they can flourish in an atmosphere of tolerance. A successful manager of development projects has to be able to stimulate this atmosphere, and therefore be in command of creativity techniques of the most diverse kinds and simultaneously create an atmosphere of trust. Bringing his hierarchical position to bear in this phase is deadly for the discovery of application areas that promise success.

The opportunity for a successful innovation only exists if technology-based and needs-oriented innovation come together. I definitely do not want to take part in the discussion, often carried on very dogmatically, between push– and pull theory in the introduction of innovation at this point.

It's true that we now know where we can go, but which areas promise success?

One will certainly be able to agree on the first areas that one would like to tackle, in the team with positive, goal-directed management. Priority techniques are called for; incidentally, though, one of the most important priority techniques is still the decision of the boss. We should not forget that in all the talk about and with all the necessity of teamwork.

Only now we do know where we should go, and the hardest part begins, which one generally calls *technology marketing* (Fig.13).

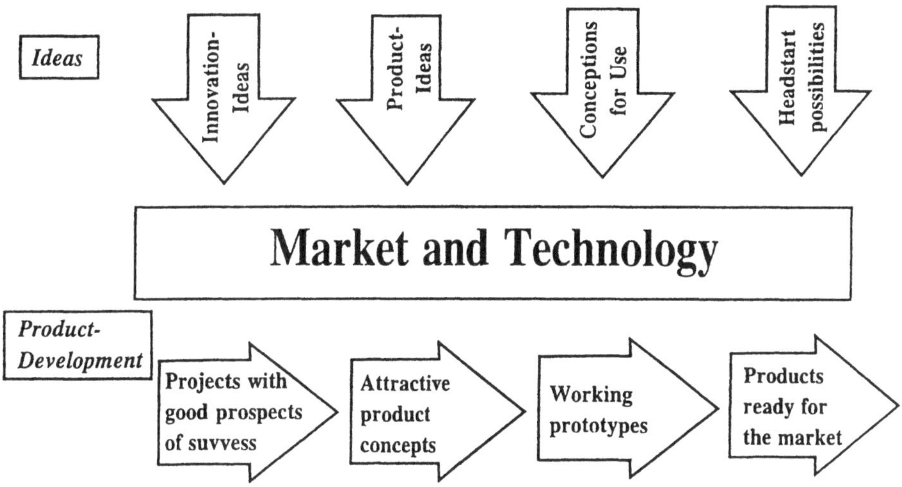

Fig. 13. From the Innovation Idea to the Market: Creative, Systematic, Goal-Directed

Technology marketing requires needs analysis and interpretation. The goal is to recognize the needs governing the demand, to push the consumer into one's consciousness and to develop new product concepts as solutions befitting the needs.

That is why technology marketing has to be go-getting and visionary and requires intensive work in persuasion.

It requires close contact with the potential consumers. In the process, though, one has to look behind the factors of the purchasing decision, which are oriented normally to the comparison of costs and performance features of the present competitive offers. Even the possibilities for technical improvement which are already being worked on, whether in one's own company or at the competitor's, do not service as prospects.

Technology marketing, on the contrary, has to ask the question: "What are the actual needs of the consumer, and which functional and economic factors bring about these needs?" Taking innovative technological possibilities as a starting point, the question has to follow: "Which new uses can we offer, and what changes result from this?"

Many successful, innovative entrepreneur admit that they carried on no market research in the start-up phase, because the potential consumers were not able to take any reliable stand on new product solutions and on innovations. They are caught up in the existing products and their applications; they know their problems, but are not aware of their real needs and cannot envision additional uses under new types of application configurations very well.

That is why the innovative enterprise has to develop the creativity and insights itself, in order to forge ahead from problem analysis of the potential users to the underlying needs. In this way, the innovative enterprise gets a thorough understanding of the consumer's needs before the potential consumers.

Two tasks result from this:

- the innovative technical possibilities have to be converted into product and service plans in such a way that the greatest-possible use arises for the potential consumers;
- the potential users have to be included in a process of cooperating in thinking and recognition which allows them to gradually get from their current awareness of the problem to the awareness of their needs and from there to the future-oriented awareness of uses.

Many new technologies are initially solutions that are still looking for their problems. That is why it is the task of technology marketing to make out problem areas and to segment according to need structures, on which the innovative enterprise can concentrate with its know-how.

Technology marketing is not the selling of y finished product, but the joint design by the developer, user and decision-maker. The joint work of all three groups is the design of attractive product concepts, working process concepts, but also innovative organizational forms.

Being able to listen and get to the bottom of things with questions, as well as creative combinations of facts are the abilities that are called for. Tools for this are value analysis, the definition of quality features, finding a price level in line with the strategy and the competitive situation, evaluation of convenience advantages, mastering interview techniques, and many other things. The creation of confidentiality agreements is frequently forgotten, though, in order to protect one's own technological position. Nevertheless, in this phase the reverse of the sentence in the first part of my remarks applies: “Silence is silver, but talk is gold”.

The developers, for the most part, assume that an autonomous build-up of the business is possible because of the actual or, alternatively, imagined uniqueness of their development. Most of the time, though, one recognizes in the course of the technology marketing process that one does not have all of the strengths available in one's environment that are needed, but that it is sensible to enter into coalitions. That is why it keeps coming to cooperation agreements, quite often in the case of young innovative enterprises, with partners that can offer production capacity and/or have a marketing and distribution network with a world-wide orientation (Fig.14).

The recognition of one's own limits is an ability that is called for in this phase of technology marketing.

Topics come up again then in connection with this such as the structuring of contracts, license structures, influences of competition law, etc.

And, for a good close, such frustrating topics for a developer as drawing up a business plan, the procurement of financing, drawing up an annual budget, the implementation of time and cost controls, etc., are then still in store for the developer.

You see that it is a long road to the real start of commercialization (Fig.15).

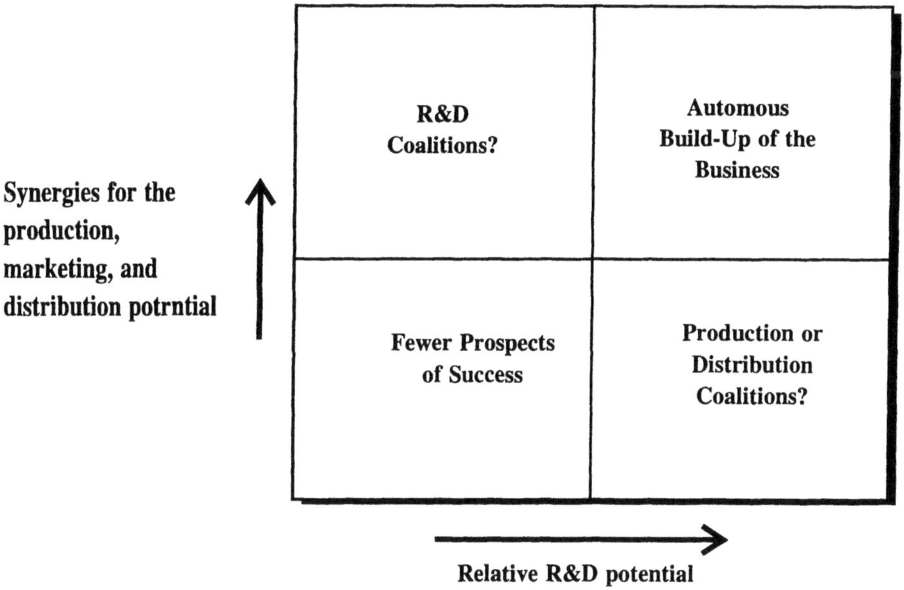

Fig. 14. Partners for Coalition ???

Steps towards commercialization	Skills	Tools
Protect know-how	Analysis Evaluation of strong points **Personel Management**	Patent research **Patent Assessment** Patent legislation
Identify market potentials	Creativity Vision **Personel Management**	Creativity techniques **Potential Assessment** Priority techniques
Identify customer needs	Customer Interviews Open up contact partners Convincing Attitude **Personel Management**	Interview design Conducting discussions Use analysis Industrial sector analysis **Feasibility Assessment** Strategic pricing Quality design
Identify strategic partnership	Analytical ability Recognize necessity **Personel Management**	Know-how safety **Weakness/strength Assessment** Licensing structures Contract designs Business plan Project Management Milestone technique
Define implementation	Design of implementation - steps - timing - finance - human resources Risks assessment **Personel Management**	**Assessment Techniques**

Fig. 15. Steps Towards Commercialisation: Skills and Tools

You already see, from the large, very diverse number of skills and tools (Fig.15), that there can be no ideal manager, but the skill will be to put together a team in which the talents for the required abilities exist.

Two abilities stand out:

1) motivational personnel management;

2) a feeling for what is feasible.

Jumping back to the Fig. 9 in which human resources were presented as a driving force for innovation, the question now arises: "Do we actually have the right people and the right corporate culture to be able to successfully implement a project?"

Depending on the life-cycle phase of the line of business, market dynamics and critical success factors, specific qualifications and employee types are required. As a result, the personnel department will also be challenged to think strategically: one must proceed systematically, in accordance with the strategic goals and demands of the enterprise, in the selection of employees and with ongoing education measures, in which targeted and planned investment should be made just like in fixed assets. The clearer the strategy is defined, and the required qualifications of the employees derived from that, the more apparent it also becomes as to which employees cannot cope with the demands or don't want to cope with them. In so doing, qualifications does not just cover education, knowledge and abilities here, but at least to the same extent the personality structure, the leadership abilities, the creativity and the style of personal relationships.

Manager types and enterprise types have to fit, and they have to meet the expectations of the strategic position of the enterprise.

We make a distinction between five enterprise types: the trailblazers, the adapters, the protectors, the balancers and the reactors (Fig.16).

1. Trailblazer	2. Adapter	3. Protector	4. Balancer	5. Reactor
• Young businessmen • New products and technogies • Willing to take risks • Management by the inner circle • Informal • Human resources constantly changing	• Builds on strengths • Clear, uniform values • Flexible, but tight • Long-term goals • Unity through "brainwashing" • Resolute striving for higher goals	• Formalized • Willing to take risks • Cost-oriented • United management group • Influencial staff	• Matrix or divisional organization • Cautious product development • Different types of managers • Market-oriented • Can only be guided over longer periods of time	• No distinct character • Imitates

Fig. 16. The Five Enterprise Types

Most enterprises have features of several corporate cultures, but a dominating type can frequently be made out. It is very significant for the success of an enterprise that it adapts its corporate culture to the strategic demands.

It is just as important that the manager of the enterprise fits the corporate culture and the strategic conditions as far as his type is concerned. Investigations show, that each type of enterprise is only capable of a few strategic alternatives. It is easier for protectors than balancers to pursue a strategy of cost leadership. Balancers are more likely in a position to find and safeguard niches than protectors. Adapters are more capable of opening up new businesses than for rationalization strategies. Protectors, in contrast, find it difficult to build up net businesses. If protectors want to expand their strategic leeway, it is more practical that they do it through the take-over of new business units or spinning off existing business units. Enterprises have to change the enterprise type in the course of their development or in dependence on the environmental conditions in order to remain viable.

Trailblazers, and that is the type of enterprise we need for the task being discussed, have to become adapters or protectors when they have reached a certain size and when their markets pass over to the maturity phase.

It also becomes clear with this realization that we will have a change in the personnel members on the way to the commercialization of a technology, and it is always an especially sensitive task for the manager to have to say to a successful developer at a certain point in time of the commercialization that it would be better for him to go back to a new development task, and that he would have to leave his dearly beloved baby to others on its path to becoming a grown-up.

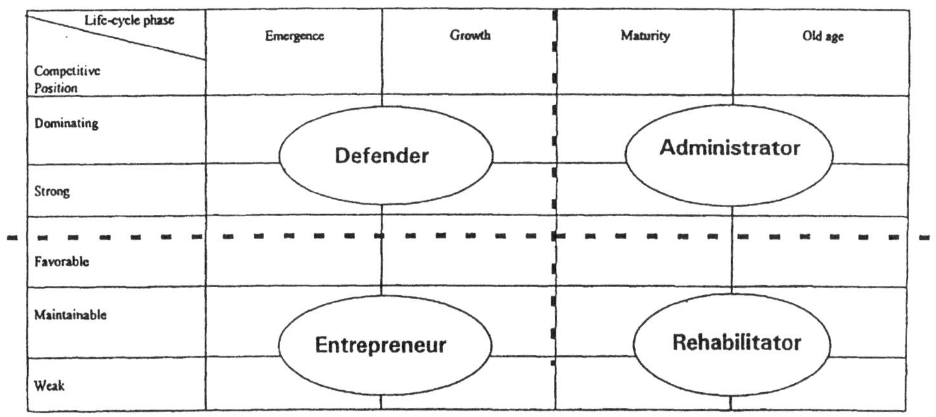

Fig. 17. Each Strategic Position Requires the Appropriate Type of Manager

Managers can also be assigned to types in their basic approach and their personality structure (Fig.17):

- Entrepreneurs,
- Defenders,
- Administrators,
- Rehabilitators.

Entrepreneurs look for the challenge of building up an enterprise with new ideas in a dynamic environment. They are convinced of success, do their personal utmost, are in a position to sweep along their employees to above-average effort and work single-mindedly to open the market, to create the breakthrough against established companies with the "right" products. Administrative tasks, hierarchical structures and routines do not interest them. Defenders have the personality to fill a strong position, radiate strength, and draw others to them because of their superiority. They love the combative conflict because they use their strengths imaginatively and quickly. They are keen on further successes, on growth and leadership. They also are not interested in administrative tasks, but they surround themselves with a strong management group, to whom they delegate important tasks and whom they inspire for their goals.

It is obvious that no manager corresponds to one of these types one hundred percent, but that several of these "souls" are united, as a rule, in one heart. Nevertheless – many discussions with managers have confirmed this, and this is corroborated in most cases with a close look – one of the types predominates in most managers.

This classification is important because each of these manager types is especially suited for specific business conditions. Enterprises that are in a weak to maintainable competitive position in an emerging or growing market need an entrepreneur as the manager. If their competitive position has already significantly improved, if they become really strong, a defender is better for them than an entrepreneur. And only a few entrepreneurs manage to switch over to defence. They more often remain entrepreneurs and continue to expose the enterprise to a stream of new ideas, fight again and again for new breakthroughs, without expanding and exhausting the position of strength that has been reached. As a result, they miss out on preparing the enterprise, to the extent possible, for the maturity phase of the business. The defenders do that better.

The fate of many enterprises lies in nothing less than being at the mercy of the entrepreneurs in their further evolution, who have led them to success and of whom no one dares to doubt whether they are still in the right place.

A correlation exists between the enterprise types and the manager types:

- Entrepreneurs are the "right" managers for the "trailblazing" type of enterprise,
- Defenders should be at the head of an enterprise of the type "adapter",
- Administrators are the types of managers the enterprise types "protector" and "balancer" need.

If agreement does not exist between enterprise types, manager types and strategic environmental conditions, this discrepancy is very quickly expressed in disturbances in the corporate culture and in a creeping loss of motivation.

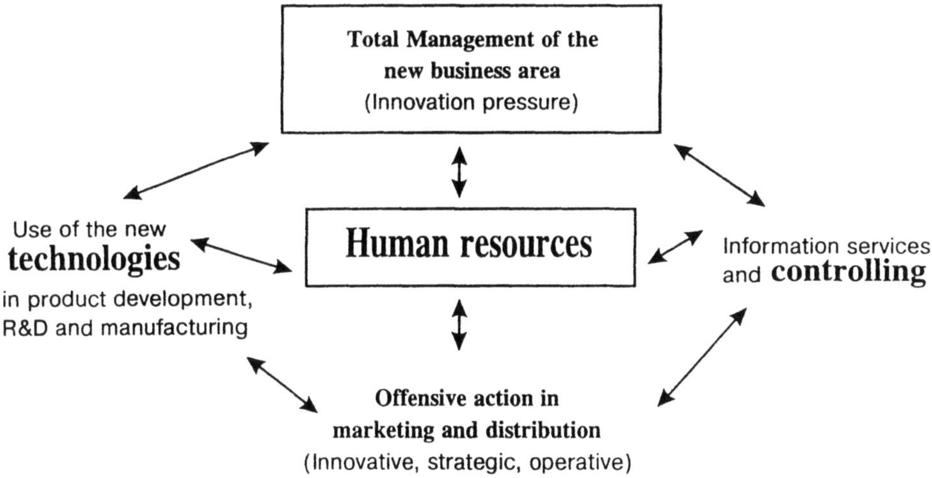

Fig. 18. Technology Marketing: The Business Structuring of a New Business

One last point as I see it (Fig.18):

The evaluation, incentive and payment system has to support the awareness of goals of the employees. They rightfully ask themselves: "what does the enterprise actually expect of me?" They require a plausible understanding of the connections between their performance and the critical results of the enterprise. This is especially applicable for executives, because their motivation structure is passed on in the enterprise. Their defined goals and evaluation criteria may not just contain individual factors such as production output, sales volume or contribution margin, but they instead have to promote a view of the entirety, That is why a part of the incentives should be coupled with the individual factors that can be influenced personally, but another part to the joint responsibility for the overall company result. The respective life-cycle phase of the business requires different evaluation and incentive criteria. In the emergence phase of businesses, the incentive system should be strongly dependent on success.

As a result, it should reward willingness to take risk and creativity above all. In the growth phase, the incentive system should be more strongly oriented to quantity.

This is a topic that is paid too little attention to, in my opinion. We should not train for innovation centres, only to determine at the end that these people drift over to the commercial structures because of the better conditions. It must be rewarding to be part of an innovation centre. The management of a new business area requires many skills and tools that are very different, but in the final analysis is the optimal deployment of the human resources in the entrepreneurial structuring of a new business, a challenging but worthwhile management task.

The shaping of the necessary abilities and the learning of the required tools is the task of the management institutes. It makes a great deal of sense to put together a team, e.g. of future trainers, in which the abilities are gathered from the point of view of principles, and to lead this team using a real-life project, step-by-step through the jungle of technology marketing to success. That is a design for training that is meant when we say. Economic Development through Management training.

Please let's not train any managers who know exactly how it goes but have not yet done it. Nothing is more stimulating and persuading for third persons as one's own experienced success.

GROWTH MANAGEMENT: THE EXPERIENCE OF SUCCESSFUL RUSSIAN ENTERPRISES

Irina SUVOROVA
Consulting-research company "ALT",
St. Petersburg, Russia

Nearly half a year ago, we, people who are professionally involved in the management industry, got quite a clear impression that among a great number of Russian industrial enterprises there appeared some which had already overcome the difficulties of the assimilation period and entered the path of further development. This group of successful enterprises is not numerous yet, however it is big enough for us to talk not just about separate cases, but about a whole phenomenon: the available potential of the economic growth at the micro-level, the level of separate enterprises.

So, we decided to investigate the phenomenon of the successful management of an enterprise during the transition period. For this purpose, we interviewed and then analyzed the results of the interviews with managers who brought success to their enterprises or managers of successful industrial enterprises.

But how could we choose such enterprises? Not at once, but we rejected the idea of using in our selection the financial results or their combinations. The reasons are as follows.

First, we all know quite well that the official data on the activities of an enterprise can be deceptive.

Second, the financial results can be indicative of success only if one considers them over a period of time much longer than the time from the start of reforms in Russia.

Third, the results of managerial decisions taken today will manifest themselves only after a certain period of time.

However, we had to select enterprises and we found the following original solution. We questioned people who had an idea of what was going on at the enterprises of the city (senior specialists of consulting, broker companies, banks of the Association of Industrial Enterprises of St. Petersburg, the Committee on Economy and Finance and the Committee on the Management of the City Administration Property, the Property Fund as well as the journalists of business newspapers and magazines) and we asked them only one question: "Which industrial enterprises in St. Petersburg can you call successful?" The success criteria were not discussed with the respondents.

We selected those enterprises which were mentioned more often.

We had meetings with the directors of 13 enterprises and tried to get answers to questions we thought of greatest importance. The enterprises we selected are given in Table 1.

C.M.R. Verkoeyen et al. (eds.), Management Training in High-Tech and R&D, 79–88.

Table 1.

The most successful enterprises of St. Petersburg (according to some specialists)

Enterprise	Kind of activity	Turnover in 1995, bln Rb	Number of employees
Ship building company "Almaz"	Building of high speed boats and hover crafts	28	1500
Joint-stock company "Baltiiskii zavod"	"Building of civilian and military ships, including ships with atomic power installations	500	7000
Joint-stock company "Baltika"	Brewery	192	850
Joint-stock company "Zvezda"	Manufacturing of diesel engines for ships, shunting diesel locomotives and trains, production of diesel generators	240	5000
Joint-stock company "Zavod Turbinnykh Lopatok"	Manufacturing of turbine blades and semi-products to use in power turbine building	125	1850
Joint-stock company "LOMO"	Manufacturing of optical-mechanical and optical-electronic equipment	146	9500
Joint-stock company "Krasnii Vyborzhets"	Non-ferrous metallurgy, rolling, drawing, casting, pressing	420	2200
Joint-stock company "Pervomaiskaya Zarya"	Clothing making and sales	20	1200
Joint-stock company "Russkie Samotsvety"	Jewelry production. Manufacturing of special equipment and research into jewelry.	420	3500
Joint-stock company "Zavod SPU"	Manufacturing of production equipment control systems	35	700
Joint-stock company "FOSP"	Men's wear production	50	1500
Joint-stock company "Khlebnii Dom"	Bread and confectionery production	125	640
TsNII TS	Technology and equipment development in ship-building. Design and reconstruction of ship-building dockyards	35	2000

Table 2.

Decision principles	Strategy elements
Establishing normal production and labour discipline; – putting everything in order; – changes in finance management	Putting everything in order and mastering the situation
Not protection against the competitor, but the orientation towards the customer's needs; – the quality of the produce is the main instrument in the struggle for the customer; – determination of the concrete, market segment; – attention to the current situation, what is happening around, with the purpose to borrow the best; – a new product which starts selling on the market must correspond to the main profile of the enterprise.	Identification of one's profile
Active policy which takes into account the partner's demands; the client is always right; – orientation towards future; consequently, the necessity to keep contacts with former customers, to interact with the trade; – the development of one's own trademark; – the same approach to all clients; no approach like "we will do better for this client, and for this one, the worse quality will do"; entering easily new markets; – any interaction with a new reality is treated as not only a way to earn money, but also a possibility to learn something new; – realization of the importance of the sales service and concrete people involved in sales; sales priority; – radical change in pricing ideology.	Organization of sales channels
replacement of the upper management by those who can work on the basis of a concrete goal; – creation of new fields for divisions and their managers to carry out independent activities; at the same time, maintenance of methods of centralized management; – use of planning mechanisms for the determination of objectives, co-ordination of measures and control.	Updating the management.
firing people who will not have enough work in the nearest time; at the same time, leaving the best; – strict approach: encouraging one who can assimilate to new conditions and firing another who cannot, no matter how useful he was before – active attraction of new people, selection of "super-personnel"; – desire to explain to people all your measures, their relation with the prospects of the enterprise the image of which is created by these people.	Development of people's self-motivation.

Probably, many managers would reluctantly agree with the fact that at the start of the planned economy collapse, **all** enterprises found themselves in the similarly difficult situation. One would rather say that the enterprises indicated in the list of successful were in a specific situation. Say, a better conjuncture due to the difference between prices of the internal and external market. Or, fixed assets renovated at the state's expense during the last years of the Soviet power. Or, the closeness of the management to the state money sources. Or something else.

For example, "Khlebnii Dom" (Bread House). This enterprise bakes bread which has always been bought and is going to be bought in future. Indeed, it sells and will sell well. However, bread consumption decreased almost by 30 % after the stop of subsidies. As a matter of fact, this bread was baked in ovens which had survived the siege of Leningrad.

Another example: the SPU plant which was built on the credit money it got from the state Foreign Trade Bank at the very end of the 80's. An excellent plant, it was built by Finns and provided with state-of-art equipment. When the plant was built, it turned out that its produce is absolutely useless, since it was intended to make digital control program systems for the Soviet machine-building plants which are known to be idle at the moment. But the plant had to pay the money back. "The plant which should not have been built", says Vladimir Stryapin about his enterprise.

In 1992, the ship-building plant "Baltiiskii zavod" (Baltic plant) was finishing the missile cruiser "Petr Velikii" (Peter the Great), the pride of the Russian Navy. It was built to the state order, at the state's expense. Then money stopped. In addition, the plant was to continue building the chemicals ship carriers under the contract signed by the former Soviet Union with the former GDR. The contract price of one ship was 12.5 mln Marks, while the price of only one set of equipment (as of the beginning of 1992) was 22 mln Marks.

The plant "Krasnii Vyborzhets" used to produce among other things coin bands. The band which was used for coining copecs and roubles. Then tens of roubles, "fifties" and "hundreds". And then, they stopped making coins in Russia at all.

The brewery "Baltika" which is famous now throughout Russia was in construction for 12 years. The equipment provided by the fraternal Czechoslovakia stayed stored during all these years at the construction site. Taymuraz Bolloev started with the replacement of rusty pipes collecting money for this purpose by small portions. And then, to get the credit for the purchase of raw material (the plant did not have money) he had to mortgage the whole plant. And even the bank which gave the credit was sure that the plant would not pay it back.

Or other examples, the plant "Zvezda" (Star) which manufactured diesel engines for navy vessels and submarines, the ship-building company "Almaz" (Diamond) which built high-speed missile boats and landing hover-crafts; "Zavod Turbinnykh Lopatok" (Turbine Blades Plant) which have been manufacturing blades for power turbines; TsNII of Ship-Building Technology etc.

The managers of any of the above-indicated enterprises could undoubtedly give thousands of objective reasons for a would-be failure. But these enterprises are a success. Why?

At this point, we are coming back to the question "Why?". Let us assume that we do not need a lot of factories which like the "Pervomayskaya Zarya" sew lady's wear. Let us also assume that the analysis of the macro-economy shows that only one factory should be left out of each ten. Just we do not need so many factories, a lot of reasons why – import etc. But which one out of ten is to be left? What does the choice depend on?

Our investigation convinces us that at the turning-point, when "old" industrial enterprises could have had their second birthday, practically everything depended on some "soft" factors. First of all, such a factor was the manager's personality and his readiness to take all the responsibilities for the future of his organization. His readiness to determine the development prospects of his enterprise in absolutely uncertain conditions and, of course, his readiness to change, if necessary, his methods and himself. Whether or not managers could use opportunities they initially had and which appeared later depended on this readiness only (this is our opinion). "Concrete people can do concrete things. Everything depends on people who have clear objectives and know what they are doing", we were told at one of the successful enterprises. And this is true.

What we saw, we called "the management oriented towards growth" or, in short, "growth management". We understand the word "growth" as a synonym of the word "development" and put some qualitative meaning into it. Let us repeat again: it is impossible to say whether or not the company is going to be viable only by the growth of the volume of sales. But if the enterprise does know its customer and is able to quickly adjust its production to meet new demands of the market, or if the enterprise has such a system of financial planning which allows it to reasonably control finances, in brief, if the enterprise has got the features it had never had before, one can say that this enterprise is developing.

The "growth", according to our understanding, means that the enterprise has got new features it had not had before, but these features will help it to work today, tomorrow and the day after tomorrow. "Now we start thinking why we work here, why we come here each day. What is the point in coming here? At first we thought of earnings, the worthy life. And now I understand that the most important thing is the survival of our factory by any means in any conditions", said Galina Sintsova. If somebody thinks it naïve, we would like to remind you that she is the director of the factory "Pervomayskaya Zarya", the factory which sews lady's wear, employs 1200 people and which, in spite of all, exists, opens its shops, have its employees trained abroad and sews not so bad clothes. The knowledge of one's customer, new channels for sales, trained personnel, these are the new features "Pervomayskaya Zarya" did not have several years ago. This is what we call growth.

When a manager says that one day he realized that the word "to survive" must disappear from our dictionary, that "it is already too late to survive, one has to work now", as we were told by Oleg Shulyakovskii, director of "Baltiiskii Zavod", in our understanding, it is also an evidence of growth. Because the reluctance "just to survive" means that the organization has got the most important feature, the desire not to adapt to the new reality, but to consider the external environment only as the conditions in which the organization has to develop.

It is characteristic that none of the managers we spoke to did not talk of their goals as of some figures they wanted to achieve.

"This is the factory which has a goal to sew good suits", that is how the managers of FOSP say about their enterprise. And they add: "The very production is to be continuously improved, now and in future". "Our plant must correspond to the

European standard, both from the inside and the outside", these are Valerii Chernyshov's words. "I would be happy if women could proudly buy our clothes", says Galina Sintsova. "Our corporation can solve all problems concerning jewelry", says Alexander Gorynya.

When talking about their goals, "successful" managers operate with concrete images and not with abstract financial indicators. But these goals require that the enterprise should acquire new features it has never had before. They do not operate with figures, because behind these concrete goal-images there are concrete steps to be taken. For example, the European standard, according to Valerii Chernyshov, is a plant of the world standard, which can manufacture produce for the largest power machine-building companies of the world. And he realizes that they have to do many thing to achieve this level and it is important that he knows what is to be done.

The essence of the "growth management", according to our definition, lies in the development by an organization of new features. Since any changes represent a long and difficult process, the growth management is directed at a long perspective. As a result, the work of the manager himself undergoes principal changes: it is not just a reaction to a situation and even not an attempt to predict future, it is some building of the future.

Only when a manager has some vision of the future, when he knows in which direction to go and this knowledge is also the knowledge of what is to be done tomorrow, one can say that it is the growth management.

Five features of management oriented towards growth

The universal "success formula" is impossible to derive. This is obvious. But talking with the directors of successful enterprises and later on re-reading the texts of the interviews, we concluded that all of them, telling us about what and how they had been doing things during the last years, said about one and the same thing.

What they have in common lies not in concrete measures determined by the specifics of each enterprise. No, it is the common business philosophy, a set of basic principles which managers use to develop their strategy. These principles help them to solve current problems. The features of the "growth management" as we determine them are as follows:

- vision of prospects;
- hope for one's own forces;
- turning circumstances into opportunities;
- orientation towards concrete steps;
- concentration on the most important.

1. Vision of prospects

"I try to sit on the cold ground all the time. I try to raise myself and look at everything from above". Actually, we could finish all explanations in this part of our report by this quotation from the interview with Galina Sintsova. Everything what is done by successful enterprises is incorporated into their idea of the future of their enterprise.

It is not a strategy, it is some vision. Because it is not a strict sequence of steps, but an action in a chosen direction, when one is ready to compare one's course with what is going on outside and to change, if necessary, the strategy, but not the direction. Using the expression of Herbert Simon, psychologist, Nobel prize-winner, all this reminds one of an ant's track projected onto the paper. If to look at this projection, one can see not a direct, but a broken line consisting of numerous segments, loops and dead-ends. But this broken line leads to the ant-hill. The same is with "successful" managers; always comparing their steps with their image of future, sometimes making mistakes, they lead their company in the chosen direction.

All steps made by "successful" managers are in agreement with their vision of future. "I think that a week without stops is a local task which can be solved in the process of the current work", says Valerii Fedorenko. "The changes in the personnel resulted from the working objectives". "Why are we known? Because, however difficult it was, we sewed and promoted our "Pervomayskaya" clothes", said Galina Sintsova.

The vision of prospects is the reason why successful enterprises, even during the most difficult time from the financial point of view, continued financing scientific research, whereas many enterprises reduced the scientific personnel in the first place. "Why are we successful? We switched our brains on. Usually, enterprises get rid of science, of brains. As for us, we decided to leave the whole scientific-research division of our enterprise and it helped us a lot at a later stage", said Alexander Gorynya.

Naturally, the vision of prospects does not appear at once. But, as it was mentioned by Taymuraz Bolloev, "there always exists some basement, some ignition energy, some starting impact". At "FOSP" for example, such a starting impact happened in early 1992, when an unknown Swedish bought a batch of suits at a price which was rather good even for Europe, 60 USD per piece. This Swedish has never appeared again and what he had done with the suits is unknown. But this was the moment when they started think about the perspectives at "FOSP".

Probably, all enterprises had similar situations. But they did not help everybody to see the future.

2. Hope for one's own forces

If you are told that they do not know how to prepare a business-plan in Russia, do not believe it! One of our clients, during several years, was making and re-making business-plans to submit to a potential investor. At first, it was a business-plan which met the EBRD requirements, then the requirements of one of Russian banks, then another and so on. From a matrix printer he went over to a laser one, from Russian paper to Finnish. The number of pages went down and the number of plots and diagrams went up. However, the enterprise was doing worse and worse. May be now, our former client is developing a plan of financial normalization in accordance with the requirements of the Federal Board on Bankruptcy of Enterprises.

When Valerii Fedorenko, director of "Khlebnii Dom", realized that it was necessary to reconstruct ovens, he also started with a business-plan. But where he could find money? It was problematic to take it from a bank, the interest rates were so high. "We thought of it all the time: where to take money from, how to use it, how to reduce expenses. Then we learnt how to do business. Money will add to money. When one learns to add money, it appears at once."

This rule "money adds money" works perfectly. And successful enterprises, unlike others, realized it very quickly: if you need money, the last thing you should do is to look for investors. Nobody will give money to somebody who is not doing well. Therefore, they preferred to gather internal resources and reserves.

When in 1992, the payment system had completely changed, "Russkie Samotsvety" (Russian Gems) found themselves between two fires. Metal suppliers required advanced payment, shops would pay only on having sold the goods. According to Alexander Gorynya, the enterprise was in a situation when there was no capital flow. One could accuse Gaidar of every failure. Some of those who accused him disappeared, others reduced the production volume and the number of employees by 4 to 5 times. As for "Samotsvety", they gave shops the permission for a delay, it was impossible to do otherwise, and started looking for the ways to find metal. They travelled across the whole Russia, reached Magadan and Chita, started using scrap, tried to buy gold from people, exchanged their produce for the metal, did their best, even set up their own metallurgical plant. And these activities which were on the edge of hopelessness helped the enterprise to survive. "The most important in such a case is the work of your brains", says Alexander Gorynya. And the hope only for your own forces, we would add.

Being very sceptical to the government, taxation etc., none of "successful managers" did not put these problems among the most important ones. "We can weep and say: this one is bad, that one is also bad, but the matter is that one must work hard, just do his dull, everyday work, no other combinations", this internal attitude has been developed by the managers of successful enterprises who try to teach their employees this approach. "I explain to my employees: there is nobody who could just solve and change everything at once. We have to deal with our current, small problems", says Galina Sintsova.

"Successful" managers are absolutely realistic and constructive in their understanding of the situation. They do not wait for a miracle or some panacea from all mischiefs. They do their work hoping only for their own forces. They do it and win.

3. Turning circumstances into opportunities

Turning circumstances into opportunities is one of the most characteristic feature of "successful" managers.

1500 seamstresses who worked at FOSP knew only how to sew "Soviet" jackets. This circumstance which could in any other situation serve as a justification of idleness was turned into an opportunity: the same 1500 seamstresses work at the enterprise and work perfectly and it is managers who have to think of competitiveness, to provide their employees with necessary patterns, equipment and technologies.

During the Soviet time, there were 6 ovens for bread baking at "Khlebnii Dom". After the bread subsidies had stopped, only 4 could work to full capacity. On the one hand, expenses on the maintenance of the idle equipment which enhanced the financial burden of the enterprise, the situation many managers know so well. And on the other hand… "We realized that our reserves could turn out of no use. Realized that we could use only 70% of capacities, removing at the same time some production lines from operation, installing new equipment", this was how they changed the situation at "Khlebnii Dom".

At the "SPU" plant, they built a big shop to manufacture various closets and stands for measuring equipment and digital control systems. Too big a shop for the number of orders they had. At first, they started welding frames for furniture in this shop (it was the very first order from the Cherepovets metallurgical plant which later on turned into the biggest customer of the plant). Then they started making beer and milk stalls, militia booths, ad posters, phone booths, and in the end they had a source of constant and quite good income.

"In St. Petersburg, we live in the close vicinity to Europe", this geographical circumstance is considered by the managers of the ship-building company "Almaz" as a strong point of their enterprise, since hundreds of ship-building specialists come to St. Petersburg from all over the world and each of the visitors may become a source of valuable information on what is happening in the world, on new tendencies in ship-building and, on the other hand, may see and tell everybody about the potential of "Almaz".

4. Orientation towards concrete steps

It turns out that "successful" managers do nothing other people cannot think of. But there is one principal difference: they prefer concrete measures.

Doubtless, they have a profound analysis or research made, if necessary. But this does not replace concrete actions. In uncertain conditions, one can be busy with the improvement of the forecast accuracy for a long time. Successful managers do things and learn from the results of their actions.

"You know, I am not used to think of problems for a long time. They are to be solved. It was not very difficult. Just it was a period when we changed our profile", says Alexander Gorynya. "We all like scandals. It is much more difficult to create".

If "successful" managers come to a conclusion that something helps or hinders the development of an enterprise in the chosen direction, this realization immediately turns into a plan of concrete actions, which starts working at once. When at the Plant of Turbine Blades they realized that they had to fire people, they did it, although it was very difficult from all points of view. "The problem is in the realization of whether or not this or that is necessary. And this is all", says Valerii Chernyshev. "Realize and act!", this is the slogan of "successful" managers.

"If you know what is necessary to improve the quality, but you only declare that your quality is low and explain why and you do nothing to improve it, you only state the problem. Another way is when you find people who agree to work, because any achievement has a concrete person behind it. If somebody can do something, then this something becomes possible. There is nothing impossible", says Vladimir Mikheev. Yes, nothing is impossible. And successful enterprises prove it each day. Prove to themselves and to everybody.

5. Concentration on the most important

Managers at the head of successful enterprises learnt to focus their efforts on the main things.

Realizing that there exist things which are of greatest importance for the future of their enterprise, "successful" managers have learnt to use all the available resources for the solution of a concrete problem they consider as №1 at the current moment.

"It seems to me that we were called successful because during the last five years of the existence of our enterprise we did what was really necessary, what was suggested by the time, what was to be done at this or that moment", these are Taymuraz Bolloev's words.

If the director of the Turbine Blades Plant thinks that the most important thing now is the fulfillment of the export order, since the market of turbine blades is difficult to enter, he checks and controls everything related with this order. "I check each telegram sent abroad, tens of them per week. Just not to miss some silly phrase we can write out of our Russian simplicity", says Chernyshev.

We can clearly see the steps taken by Valerii Fedorenko at Khlebnii Dom: discipline, replacement of old-fashioned ovens, introduction of new kinds of goods, competition. These steps were not pre-determined, they were suggested by life. But each time, he could determine the most important task and focus all efforts on it.

"If to work over a problem each week, the problem will be solved", said Galina Sintsova. One cannot say better.

FROM KNOW-HOW TO KNOW-WHY

Cem ERGÜN
AREND Industrial Consultancy, TURKEY

1. Change management

1.1. CRISIS = DANGER + OPPORTUNITY

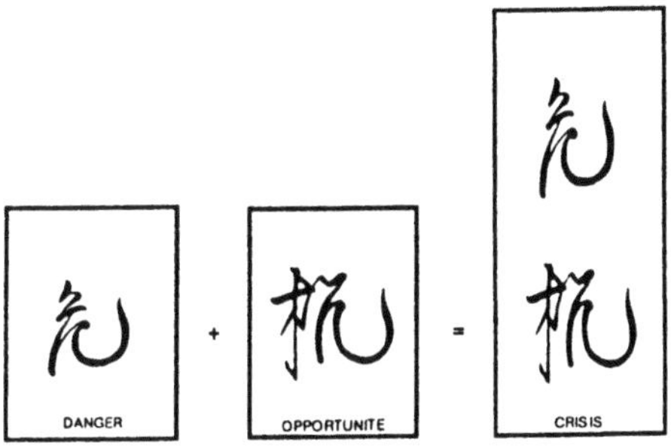

1.2. COMPANY REQUIREMENTS AND MARKET NEEDS

In unified markets of the world borderless to political and economic events, where competition among nations and among brands overlap in industry and trade and where the number of companies:

- who develop the fittest product the earliest;
- who produce fast, efficient and with quality;
- who sell at most convenient locations at most reasonable prices;
- who continuously satisfy their customers with best service,

kept increasing, the key word of 90's has been "change".

Restructuring of companies, development of corporate systems, establishment of new job designs and communication schemes, improving performance of people, creation of quality and productivity awareness, determination to prosper as an innovative, competitive world class company with a dynamic operation are issues which are gaining momentum.

C.M.R. Verkoeyen et al. (eds.), Management Training in High-Tech and R&D, 89–93.

However it has been very little emphasized that, this transition which is termed as "change" and instituted through "restructuring", is in fact a "*planned change*".

The existence of "*strategic plans*" developed with great efforts through investing into "human resources, technology and information systems" behind the "re-engineered" successful companies of today has been generally overlooked.

Benefiting from the accumulated experience, the enterprises which are lagging behind have to accommodate to this concept, utilize latest techniques and conduct their development in the light of these approaches upon structuring around their technology mission. Holding up through the remaining couple of years ahead and to be among those successful companies in 2000 depends first on a *commitment* and then on emphasizing all steps in planning and executing the change, with patience.

1.3. FROM KNOW-HOW TO KNOW-WHY

The Director of a Research and Development Organization in a High-Tech Enterprise, usually a scientist or an engineer, is considered to be the top person in charge of the "Know-How" side of the business.

The R&D Group people, with different technical backgrounds and expertise in various fields, are regarded as the bearers of valuable information of the company, relating to products and processes.

Furthermore, the specialists in the R&D function are expected to enrich the existing intellectual heritage and innovate new products, better processes and most important, additional knowledge for a superior "Know-How".

It has been long discussed that occasionally the members of the R&D community may not have the same drive with the other functions or their interests may differ from the end results sought by higher management. It is frequent that the goals in R&D are criticized for not being part of the bigger picture, commonly planned by strategists and designed by Marketing officials. Sometimes, the R&D output may not match the present necessities within the enterprise and the future needs of the marketplace.

It is also true that in some cases the mission set forth in directing the technology and the values created by the R&D effort may not be shared within the company. Though, they may reflect the unanticipated urgencies in term of strategies and the yet unfamiliar choices of customers.

To the contrary of a widespread belief, the realism in anticipation, the spirit of change, a much wider understanding of the business perspectives and the concern for the wealth of society are inherent to the very nature of Research and Development activity.

The scientist, the engineer, the researcher may have a deeper sense of imperatives but may occasionally lack the wisdom to express them in strategic terms.

The role of the R&D Director becomes critical in such cases where a fundamental shift, from "Know-How" to "Know-Why" is required within the R&D group, to promote the participation of its members at maximum capacity of their full potential.

Another crucial point in this role is the ability to influence the change in the company, to induce the contribution of all others, and lead the transition to be through a "planned" path, focused on technological strengths.

2. Leadership in directing R&D

2.1. LEADERSHIP

While often recognized with its technical ability, a Research and Development Director is also a highly capable person with a rare combination of talents, education, experience and interests. In addition, the job requires to be an administrator, a businessman, or a salesman at times. Such a diversity form a set of necessary conditions to be appointed but this is not sufficient for success. Because, leadership is by far the most important trait in an R&D Director's profile and determines the extent to which he or she is accepted by the rest of the organization. This should be an esteem at corporate level and beyond the appreciation for daily coaching of teams in managing the R&D.

As intended, leadership is power, leadership is initiative, leadership is responsibility.

In difficult times, and nowadays this means at all times in high-tech businesses, the leadership of the R&D Director is rather the key tool for the overall success. There can be no leadership without a mission and there is no mission accomplishment without a leader. The mission conceived and set forth by the R&D Director may provide the intuition for himself as well as results for the enterprise.

2.2. TECHNOLOGY MISSION AND STRATEGIC PLAN

The leadership of a Research and Development Director becomes vital in such periods of transition, especially when the necessity is imposed by changing external conditions.

This leadership is to be an action to redefine the technology mission for the R&D Group, with contribution from the key people within group. But it should progress further into an involvement with the strategic plan to incorporate that mission as the essence of the enterprise's excellence and destiny and the readiness to take challenges to overcome the inertia of the change.

It is a strong conviction that for all high-tech enterprises, the technology mission should form the core of the company's strategic plan. Especially for small or medium size operations, this mission should be based on competitive product development.

The motto "Create today, produce tomorrow, sell the next day and make profit" should be the basic approach for product innovation and improvement and shared by all parties involved. The R&D Group, while continuing to accumulate knowledge, pursuing latest information and following new technologies, should concentrate more on practical application. The focus should be on implementing the up-to-date developments for all new investments and for modernization of existing processes. The group should enhance the in-house expertise and means to develop new products, to design new processes and to engineer new plants. The people in R&D should absorb the "Know-Why" along with the "Know-How", re-adapt themselves for serving the customer needs and for providing value to improve the quality of life. Within the enterprise, they should establish the understanding to cooperate technically and internationally.

2.3. PERSONAL ATTITUDE

Along with the technical ability and leadership as the two main traits, the personal attitude of an R&D Director is also a distinctive factor for effectiveness in achieving results. Although many of them may seem to be attributes of leadership, a fair number of personal characteristics should also include the following: Intellectual honesty, Loyalty, Fairness, Conscientiousness, Dependability, Organizational ability, Good memory, Good judgment, Thoroughness, Accuracy, Imagination. As a role model, the R&D Director should be inspiring, enthusiastic, courageous. He or she should be productive, eager to learn and develop, must think, plan and act.

3. Management skills and tools

3.1. EXPANSION OF SPIRIT THROUGH A LEARNING ORGANIZATION

It is to be considered that the best way of integrating the benefits from the R&D as a resource to the rest of the organization, is to utilize the same spirit as related to other functions in the enterprise.

This may be achieved in strategic planning and management of all the operational and business issues of the company, both short term and long term, through:

- a systematic information and communication,
- an open exchange of ideas,
- an interactive responsibility sharing,
- an effective coordination of functions,
- a participative decision making

in promoting

- customer minded thinking,
- company business awareness,
- application oriented improvements,
- motivation for competitive objectives,
- internationally cooperative directions

for the success of the enterprise as a whole.

3.2. MANAGEMENT TRAINING

While the "Know-How" may gain momentum into a technology mission and as the "Know-Why" may become a shared corporate attitude, the dynamics of change and the management of transition will require additional skills.

The existing readiness for enhancement in an R&D Group can be an asset and help to accelerate the learning. The present awareness may serve as a ground to provide the R&D Director or his or her key people, with the missing management tools through a "look and learn" approach and relatively in a short period.

This approach can be termed as “Read-How” to illustrate the flow of information, via books, papers, Internet or workshops as such. Similarly, it can be called as “See-How” to describe the help of company visits to foreign countries or of orientation of managers in international exchange programs. The last but very effective approach can be named the “Show-How”, where outside support in form of temporary management assistance for on the job training or better in form of in-place consultancy through experts in both fields, i.e. in management of technology and in technology of management.

MANAGEMENT SKILLS FOR A GLOBAL ECONOMY: THE CHALLENGE OF TRANSITION

Gisela Meyer ESCOE
University of Cincinnati, USA

There is a growing recognition throughout firms in all industrialized economies that today's global competitive business environment requires certain general broad-based managerial skills in addition to numerous job-specific or firm-specific skills. As evidence of this trend consider the following:

- A number of senior executives of Fortune 500 companies have played musical chairs, moving from one industry to another. For example, consider Procter and Gamble's retired president and CEO, John Smale's move to General Motors or Louis Gestner's move from Pepsico to IBM.
- Many business schools are changing their curriculums to include a more multi-disciplinary approach (such as collaborations with engineering, education, economics, and information technology). Many too, are increasing the number of liberal arts courses they include while decreasing the quantitative ones. The Harvard MBA program has turned toward general problem solving skills and played down a number of discipline specific requirements.
- At my own institution (the University of Cincinnati, a large state university in Ohio which enrolls some 37 000 students), business school enrollments have been falling for the last five years or so. However, two new degree programs, a bachelors and a master's degree in business economics are proving quite successful. The degrees offer the best of both worlds: a liberal arts background and specific business training in fields such as marketing, accounting, and finance.

In this era of increased local and global competition, this new emphasis on managerial training seems natural. As the proportion of the workforce with post-secondary education in the industrialized world continues to rise, the business degree must do much more than serve as a simple screening device which identifies high quality students. Students are looking for degrees that make them ready for the job market. Business schools, in turn, are taking a good hard look at what firms are demanding. They note, in this era of deregulation and growing free trade, management structures must be lean and productive. A firm simply cannot afford a heavy bureaucratic structure. The distinction between management and worker must be diminished. Further, since firms frequently produce output that cuts across several product lines, knowledge tied to anyone area is less valuable.

C.M.R. Verkoeyen et al. (eds.), Management Training in High-Tech and R&D, 95–99.

Training the generalist manager is especially challenging in Russia, where management's structure has been extremely hierarchical and overwhelmingly technical with a high division of labor. The sort of broad based humanistic background that today's premier business schools are emphasizing is well outside the traditional Soviet approach. There have been a multitude of post-reform training efforts that range in approach and quality but most of the programs have attempted to deal in some way with this general challenge.

This paper will focus in detail on five of the most relevant general managerial skills absolutely necessary for success in transitional Russia. They are: communication, priority setting, realism, making a Win/Win framework, and mitigating risk and uncertainty.

1. Communication

Communication is two-way. It is absolutely essential that one understands another, as well as be understood. Effective communication requires much more than a common language (and often is quite possible in a multilingual setting with effective translation). First, and foremost, it requires effective listening and understanding. Next, it requires a standardization of terms measurement units, currencies, accounting practices. Russia has seen significant progress at this level. Numerous programs are currently underway to train personnel in Western accountancy. A new initiative by the U.S. Commerce Departments National Institute of Standards and Technology's will provide a comprehensive standards training programs for engineers, administrators, and technical experts from the Newly Independent States ("Business America", 1995).

Furthermore, effective communication requires the ability to communicate both orally (in daily conversations, on the phone, and in presentations), and in a written framework (in memoranda, business letters, and contracts). Finally, while many of these skills can be taught, the most eloquent of speeches or the slickest of presentations will fail to achieve its communications objectives unless it can be believed. Thus, effective communication requires trust that comes from understanding and respect.

2. Priority setting

Effective management requires a manager to have a clear vision in mind. He or she must establish what is "job one" and follow through with this goal. Ranking ones objectives requires a thorough and complete understanding of the enterprises' mission, goals, and long-term growth strategy. Determining the best way to achieve a goal requires a manager to evaluate the relative costs and payoffs of different strategies. This is a particularly difficult task in a transitional environment were economic and legal conditions are changing rapidly. This sort of situation requires that a manger not only be able to establish and pursue priorities but that (s)he is flexible enough to adjust one's goals appropriately. The rigidities of a formal plan must be abandoned. In its place there must be a dynamic strategy that evolves as conditions change.

Next, the successful manger will develop strategies for sharing this information as needed with the employees so that everyone can work as a team to meet the given objective. If a team member is uninformed of a goal and because of this, an objective is not achieved, the manager should be accountable. A manager must effectively communicate company goals, lay out a plan of implementation, which includes individual goals, and check co-workers' progress toward its achievement.

3. Realism

A Russian manger must be able to look at his or her own situation with a critical eye. It continues to be a fact that a significant percentage of Russian production is actually value-subtracting (estimated at 8 % of total output, Warner, 1994). This problem is especially true in the defense sector were improvements in international relations has meant decreased demand for weapons. Other high-tech industries are a mixture of high quality and obsolete plant and equipment. Management must be able to evaluate the quality of the equipment at hand, establish what is of adequate quality, what must be replaced and what should simply be scrapped. I, and numerous other economists, suspect that much too much effort is being put into trying to gain ownership rights of technologically obsolescent production facilities. Anne Krueger (1992) points out, the experience of developing economies suggests that growth takes place primarily through the emergence of new activities, not through the adoption of older ones. Existing productive assets in Russia are frequently based on outdated technologies, utilizing too much capital, and they are of an inappropriate scale. Given the immense human capital of the Russian worker, a comparative advantage in high tech industries will best be achieved through new investment (either foreign or domestic).

4. Making a Win/Win framework

The successful manager must understand that a competitive environment is not ruthless but suggests a win/win framework. Business transactions take place in a market economy when both the buyer and the seller perceive a gain is to be made by trading. Thus competition requires that both demanders and sellers are in a position to gain from doing business with each other. Profits occur because a firm can successfully supply a consumer with what (s)he desires at a price that is lower than the consumer is willing to pay for the good or service. Thus, managers must be acutely aware of the needs of their potential business partners and costumers. It must be their goal to meet these needs to attract business. This requires a firm to undertake market research. For high-tech industries, in which there are only a few major customers, this may best be achieved by plant visits, product demonstrations, and other such activities. Management must be extraverted and make an effort to know his or her consumers. This same approach is necessary in setting up international joint ventures. The successful firm will have a plan that is attractive not only to their own bottom-line, but to their partner's as well. These arrangements, then, must be formalized into your company's objectives. What is good for your customers and for your partners is good for you!

Managers should also keep this principle in mind when dealing with employees. Even in the turbulent transitional labor market high quality employees are swept away by competitors if they are not adequately rewarded. Similarly, the least productive employees must be helped to improve their productivity or they must be dismissed and most often replaced. Such high labor turnover can be very expensive for a firm. Thus, profits will accrue to the firm that appropriately motivates, trains, and rewards it's workers, but is still willing to let someone go if they are ineffective.

5. Mitigating risk and uncertainty

Although by far the largest of the transition economies, Russia has failed to attract its share of foreign investment and joint ventures. This is primarily the result of political and economic instability. Even though the political situation has become somewhat clearer with the reelection of Yelstin, the constantly changing, and capriciously enforced commercial legal environment will remain an issue. The pervasiveness of black-market activity (especially in transportation) also remains a serious obstacle to foreign investment. Western business people are extremely uncomfortable in this sort of environment and often choose to look for a business relationship in the more predictable, although arguably less promising markets of countries such as Poland, Hungary, Slovenia, the Czech and Slovak Republics, and Estonia. The successful Russian manager will recognize this situation and must undertake the difficult task of mitigating this uncertainty.

Clear, truthful, and frank communication is the first step. Next, since it is quite likely that the Russian firm has a higher tolerance for risk than does its Western partner, the Russian firm would do well to act as a buffer. This can be accomplished by providing a guaranteed maximum price for delivered goods. For example, one business I have consulted for, backed out of a major deal with a Russian firm because they suspected that during delivery of their cargo to the port, they would likely have to pay extortion money. They were entirely inexperienced with having to deal with this sort of threat, and found it unacceptable. They decided to stop pursuing a very sizeable business deal because of this situation. If their Russian partner had guaranteed shipment of the product to the port on a given date for a prearranged price (thus internalizing the black-market costs) the deal would have gone through. Failure to appropriately handle risk botched a very big business deal.

Other creative approaches that might be used to mitigate risk include risk pooling (through private insurance for example), and the use of escrow accounts to guarantee payments. Furthermore, Russian business should become an active lobby against organized crime. The continued quiet toleration of such practices is bad for business. It makes Russian firms less attractive than her foreign rivals by raising the perceived cost of doing business with Russia.

6. Conclusion

The five skills outlined above are not meant to be an exhaustive list of necessary competencies. Rather, they represent a careful examination of those issues which are most relevant to the transition situation in Russia today. They represent the sort of broad based, humanistic skills necessary for a successfully functioning business in any country. The ever increasing global nature of business today has resulted in a redefining of business training. This is not to the exclusion of the technical basics included in any well rounded business programs (such as economic theory, statistics, accounting, finance, marketing, organizational behavior, and international economics). But, rather, this process has resulted in the expansion of this list to include general considerations like those discussed above. Business acumen requires more than technical proficiency. Successful management is about motivating colleagues, innovating product lines and production methods, intuition into ones customers, and an understanding of both your own and your rivals' strengths and weaknesses. These sorts of characteristics transcend wrote learning.

As major players in Russia's high technology industries, you are cursed and blessed to be living in exciting times. The challenges are extreme and the rewards are immense. I wish you luck, success, and prosperity.

References

1. *Business America*, "U.S. will provide standards training to Russian and NIS", V. 116 № 1, 1995, p.10.
2. Krueger, Anne, "Institutions for the New Private Sector", *In The Emergence of Market Economies in Eastern Europe*, Ed. Christopher Clague and Gordon Rausser, Cambridge Ma: Blackwell Publishing, 1992.
3. Warner, Malcolm, "How Russian Managers Learn". *Journal of General Management* V. 19, № 4, 1994, pp. 69–87.

AN INTEGRATED APPROACH OF BUSINESS DEVELOPMENT AND INFORMATION SYSTEMS DEVELOPMENT

Herman L. LEIJDESDORFF
Business Consultant Cap Volmac (NL).

More and more information technology (IT) becomes a strategic factor in the development of existing and new businesses. The great impact of IT on the success or failure of an organisation requires an integrated approach to information policy, where the starting point is the organisation, its strong and weak points, its strategy, its environment.

Information policy, the base for systems development, is able to reinforce the strong points of an organisation and to take advantage of the opportunities offered by the market. Recent developments in information technology make such developments more and more challenging. Almost daily we read about companies which succeed in obtaining a competitive distinction by the use of IT. No longer IT contributes only to cost reduction or raise of effectiveness or efficiency, but grace to IT new services, new type of business and new quality levels become possible.

In the beginning, the 60's, IT was hardware orientated and focused on administrative application. This led to a situation of "island automation" which caused lots of control and management problems in the 70's. As a reaction in the 80's IT attempted to improve its organisational structures. Finally, in the 90's IT become more mature and the accent is more and more put on business improvement and contribution to business strategy.

The consequence of this development is that organisations become more and more dependent of the use of IT and in some cases one may say that the survival of a company even depends on IT (e.g. banking or airline reservation systems). This means that information technology becomes a field of attention for general management. It is the responsibility for general management to make the right choices in this so fast developing technological world. This requires, on one hand, a fair knowledge of topical IT developments and, on the other hand, insight in the effects and consequences for the organisation and the business. An information policy, as a result of the general company policy, should be set for the entire organisation. And this process should be elaborated in close cooperation with the general management.

Here a method will be described to facilitate strategic decision-making in information supply. This leads to a policy not reacting to short term problems or technical issues but anticipating real issues which challenge the organisation and the business. On the one hand external triggers are to be considered, like customer's demands, competitors' behaviour or suppliers' opportunities. On other hand, the internal capabilities of the organisation must be taken into account, the strong and weak

C.M.R. Verkoeyen et al. (eds.), Management Training in High-Tech and R&D, 101–106.

points, like quality of processes, products and services, the cost-effectiveness and the time when new products or services should be introduced to the market.

The main elements of the information policy making a process are presented in the next schema.

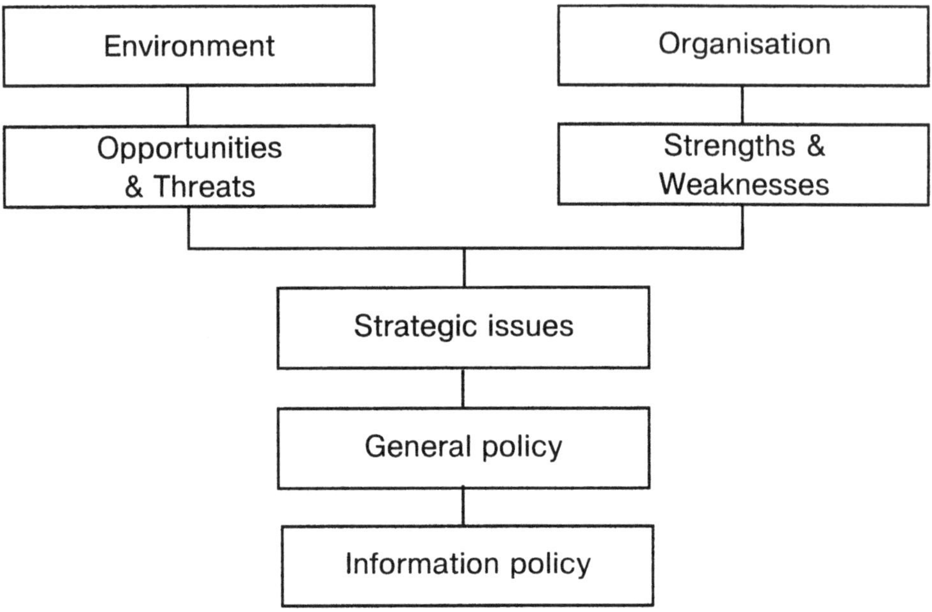

A company's information policy will be elaborated with in a process divided into five rather logical, than consecutive stages. Some stages might be elaborated in parallel. The stages are:

1 – Orientation;

2 – Drawing-up and Analysis;

3 – Strategy setting;

4 – Priority setting;

5 – Formulating information policy.

Now these five stages will be discussed in more detail.

1. Orientation.

The objective of this stage is to discover what is going on and to define the starting point and the boundaries for the following stages, leading to the corporate information policy. The main issues to be made clear in this stage are as follows:

- What is the reason for this organisation to exist? What is the added value to the community?
- What are the major objectives of the organisation?
- What products and services are offered?
- What role do these products or services fulfill for the company's customers?
- What unique "selling points" does the organisation have?

2. Drawing–up and Analysis.

The analysis stage consists of two parts: the external analysis of the environment of the organisation and the internal analysis of the organisation itself. The external environment consist of customers, suppliers and competitors, but also government, unions, shareholders, labour-market and other. The objective of the external analysis is to identify a limited number of opportunities and threats. Since a company is able to pay a considerable amount of attention only to a limited number of issues, it is important to concentrate on the most promising developments and tendencies. Within the process analyzing the organisation's environment some questions should be answered, like:

- Who are the end-users of the products and services?
- Who are the customers, who is the one to pay for the products?
- What are the thresholds that prevent new-comers entering the market?
- What is the current market position of the company?
- What are the market forecasts?

This analysis of the company and its prospects can be based on Porter's "competitive forces" concept. The result will be a large number of tendencies and developments. Examples of global trends that force major business changes are:

- Globalisation;
- Privatisation;
- Deregulation;
- Big Mergers & Acquisitions;
- Divestures;
- Technological enablers.

Next step is to select those of them – positive or negative – which are most relevant to the company. This will be done after an intense discussion of the issues and their impact on the organisation with the decision-makers. The final result will be an overview of most important opportunities and threats.

The objective of the *internal analysis* is to identify the major strengths and weaknesses of the organisation. These strengths and weaknesses might enable or disable the organisation to achieve it's corporate objectives. Also regarding the internal developments it is important to concentrate on a selection. By analyzing the internal functioning of an organisation the following questions should be answered:

- Which business processes directly originate from companies' objectives? A "value chain" (Porter) analysis will give a fair insight in business processes and their dependencies. A distinction is made between the primary processes and the supporting processes.
- How is the steering of processes arranged and what about the reporting by the accountables? It is a matter of keeping the planning (the one way) and the control (the other way) in balance.
- What are considered to be the critical success factors of a company? Is management indeed steering these issues?
- At what stage of the life cycle are the products or product/market combinations?

3. Setting the (information) strategy.

In a creative process the decision makers of the organisation point out an future strategy taking into account the topical position and the most likely way of development. This is the place to combine the earlier indicated strategic elements: the external opportunities and threats and internal strengths and weaknesses.

As a reply to business changes an increasing use of information technology (IT) is possible in fields like:

- Customer focus;
- Time to market / competition;
- Cost reduction / increased effectiveness;
- Survival / recession;
- New business processes / increased effectiveness;
- Quality improvement of products / services;
- Decentralisation of business processes.

The next step is to investigate in more detail to what extent IT and information systems might contribute to achieving corporate strategy. It is important to realize that this is a two-way interaction; some business needs might require a certain IT solution, but new IT developments might as well trigger new businesses.

Examples of promising "new-IT" areas are: expert systems, information warehouses, decision-support systems, groupware, Internet, e-mail, video/PC conferencing, distance training, virtual reality, process support software and workflow management, advanced multimedia, pattern recognition, electronic data interchange, robotics and many other. The more "information-intence" the business process is the bigger will be an opportunity for IT to come up with relevant solutions. In some cases it will be possible to obtain a competitive advantage or to create new products or services by introducing new IT applications. A topical example, of course, is Internet.

4. Setting priorities.

After information strategies have been pointed out at the previous stage, they will be assessed and choices will be made. There are three points of view:

– Suitability: to what extent does a strategy fits the situation and its particular demands;

– Feasibility: will it be possible to install the particular solution and bring it into practice;

– Acceptability: is the proposed solution acceptable for the organisation.

The answer to these questions will depend on, for instance, the organisation's level of expertise and experience, resources available, the attitude towards changes and towards technology, the corporate culture and management style.

The stage-theory of Nolan is a good help in elaborating this assessment from the IT point of view. Nolan identifies six stages of development in the application of information systems by an organisation. The theory says that an organisation should pass through these stages in a consecutive way. No stage should be skipped in the development of an organisation. The following stages are distinguished:

– Initiation; main objective; cost reduction;

– Expansion; the wider spread, more or less uncontrolled, of the initiation stage system;

– Formalization; improving the control and restructuring of information systems; improving the documentation;

– Integration; linking applications by means of data interfaces;

– Data-administration; an integrated approach to business, reflected an a corporate data-model;

– Maturity; reflection of corporate business processes in information systems; common use of data and information systems functionality all through the company.

This phase of setting priorities is important to reduce the risk of failures at the next stage of development.

5. Formulating the information policy.

This final stage leads to tangible objectives and indicates the way how to achieve these objectives. Also the general conditions and starting points for information systems development will be formulated.

The information policy will address the following six aspects, characteristic of all organisations:

- Information need;
- Personnel aspects and policy;
- Organisational structure;
- Financial planning; budgets;
- Technological possibilities;
- Administrative procedures.

Finally, it is good to stress that formulating a company's information policy should not be a "once in a life time" happening. It is advisable to update information policy on a regular base, once a year or once every two years. The internal and external conditions may change very quickly, the IT develops even faster. This will constantly offer numbers of opportunities, but might turn into threats if not managed in a proper way.

THE MANAGEMENT OF INTERNATIONAL COLLABORATIVE SEARCH PROJECTS AND THE IMPACT OF PARTNERSHIPS

Luigi MASSIMO
Head of Unit Commission of the European Communities

1. Introduction

The majority of papers presented in this workshop are dealing with training of full time managers in high-tech companies. In this contribution, on the basis of the experience of the European Commission, I would like to stress the need for management training also of those scientists or principal investigators who have to submit proposals for grants and coordinate project activity.

The organisation of international co-operation and the creation of partnerships among scientists and industries belonging to different European nations has been since many years the main goal of the Science and Technology policy of the European Commission.

This experience has proved that these collaborations, if properly managed, create long lasting benefits both at the scientific level and at the level of the social, economic and industrial exploitation of research.

More than 3 billion ECU per year are now being spent for collaborative research projects involving partners in different European countries. This activity started already more than 20 years ago, and evolved over the years to the present level. Although this amount is a rather important one, it represents only approximately 4 % of the total government spending for civil R&D in the Member States of the European Union. This means that, in order to be justified, these expenditures must show a particular added value which could not be otherwise obtained through actions conducted at the national level. Such added value is given by the synergetic effects of co-operation.

This same principle of creating international scientific projects has been applied in the co-operation with the countries of Central and Eastern Europe.

Two management aspects appear to be of particular importance for the scientists and industrial partners of these countries: the preparation of research proposals and the creation of international partnerships.

2. The preparation of scientific project proposals

The experience gathered by the European Commission during its co-operation with Central and Eastern Europe shows that very often good scientific proposals have to be rejected because of insufficient management and cost planning.

C.M.R. Verkoeyen et al. (eds.), Management Training in High-Tech and R&D, 107–113.

Scientific and Technological grants of the European Commission are normally attributed through open competitions after a peer evaluation conducted by independent experts. The subjects covered are specified in each call for proposals, and evaluations are based on the following criteria:

Scientific excellence which is of course an essential evaluation element.
Correspondence to the **specific goals** and scientific themes of-the call. The interest, feasibility and verifiability of the project objectives.
Innovative character of the proposed research.
The quality of **management**.

Concerning the quality of management it is essential that project proposals give a clear description of the S&T activity. A detailed workplan has to be specified including an assessment of the risk of failure. Milestones have to be specified and the research objectives have to be set as far as possible in a verifiable form, specifying how and when such verification will have to take place. It is obvious that, in the case of research projects, objectives have to be flexible and if necessary should be modified in the course of the activity. Indeed the failure to reach some objectives will not necessarily mean a failure of the project: spin-offs are sometimes more important than the original goals, but no management is possible without detailed objectives.

The workplan should be realistic and give a clear picture of the manpower and other means involved. It is important to demonstrate the adequacy of the means to the objectives.

Another important element in the evaluation of project proposals is given by the qualifications of the proponents and of their partners. This includes the equipment of the laboratories, the experience of the scientists involved, their past publications, etc.

All project proposal shall indicate its expected benefits, together with a cost-benefit assessment. It is obvious that for research projects it is not easy to make a good assessment of costs, and even more difficult to forecast future benefits. However the best possible indications should be provided in order to facilitate the assessment of the evaluators. Under this point of view the foreseen results must be specified together with plans and means of exploiting them. Industrial oriented research project proposals should include basic descriptions of the exploitation possibilities and show the intent and capability of one or more partners to be involved in exploration.

Administrative details must be provided specifying the arrangements among the project partners and financial details such as costs breakdown, overheads, labour costs, depreciation of equipment, etc.

Since all projects financed by the European Commission are collaborations involving partners from different countries, each proposal must include information on the number of partners, the credibility of partnership, particularly concerning its possible duration after the end of the project together with the scientific and administrative qualifications of the project co-ordinator.

During the development of the project and at its termination, measurable results should be shown. These include the verification of objectives and milestones already mentioned above, scientific publications, their citations and, whenever relevant, an analysis of products introduced to the market and other spin-offs.

3. The impact of partnerships

An important number of evaluation studies have been conducted in order to assess the impact of this European collaborative research. They included bibliometric and patent analyses, microeconomics and market studies, and a high number of interviews or questionnaire surveys with scientists, managers, industrialists and civil servants.

The impact of research programmes was analysed by measuring both direct and indirect effects. The former are those which arise as a consequence of the explicit objectives of the research project. They are assessed by drawing an inventory of results such as bibliomelric indicators, patents, number of products and processes, standards, etc. The indirect effects (or spin-offs) include all benefits in terms of technology, know-how, management and even image resulting from the participation in the activity.

These studies highlighted the benefits of international partnerships and at the same time showed the conditions under which the best scientific and economic results could be obtained.

These benefits can be classified as follows:

I) Scientific benefits

- Optimal utilisation of equipment:

Particularly in the case of expensive installations which cannot he available in every laboratory, the association of a number of different partners leads to a better utilisation of equipment and allows smaller laboratories to make use of facilities they could not afford otherwise. The pooling of resources allows the construction of large installations which could not be afforded even in the largest countries (e.g. nuclear fusion).

- Confrontation of different methods and pooling of best scientists.

- Integration of isolated teams of researchers.

Collaboration is a way of assuring the attainment of the minimal critical mass needed for an efficient scientific activity. Particularly in the smaller European countries there was a number of research teams which, while including good researchers, were rather isolated from the main stream of world science. Collaborative proposals and the need of participating in competitive tenders, forced them to become fully integrated partners of the most advanced scientific activities. Furthermore, they got the possibility of sharing equipment their laboratories would otherwise not have been able to afford.

The involvement in research programmes of a large number of organisations which had been hitherto relatively isolated and dispersed has gradually been transformed into networks for the exchange of knowledge and know-how.

- The effect of networks on scientific production.

A comparison of the citation pattern of scientific publications has shown that papers whose authors belong to institutions located in different countries, receive in average twice the number of citations than papers whose authors belong to a single institution. This higher number of citations occurs not only in the countries where the laboratories are located, but also in third countries. Furthermore the time delay between publication and citation is shorter for these transnationally co-authored papers. It is a clear proof of the fact that international co-operation creates a context in which better science can develop.

II) Organisational and Structural benefits

- Widening the choice of external support.

European collaborative research has helped breaking regional habits according to which any external support required by the organisations conducting research, particularly in the industrial field, was mainly chosen locally. As a general rule, industries needing research support used to address themselves exclusively to a few universities or laboratories in their own country. Through the participation in international programmes they learned to have a better overview of capacities, equipment availability and costs elsewhere in Europe.

Collaborative research also stimulated the creation of new types of partnerships. In Greece, for example, university/industry R&D collaboration was virtually non-existent before Community funded research activity. A considerable change in behaviour has taken place since.

- Management and organisational improvements.

The need of participating in international competition, the confrontation with the management methods of other industrial partners, with their research and production techniques has forced important improvements. Such improvements were not limited to the institutes of the poorer European regions, but were also evident in the most advanced laboratories and companies. These innovations concerned project management, quality control, production techniques, reliability of products, etc.

- Improvement of the level of the scientific and technical staff.

The confrontation with foreign partners has forced improvements not only in the top scientific staff but also at all support levels.

- Permanence of partnerships.

Many partnerships created by international collaborations tend to continue after the termination of the projects for which financial support was given. The 1992 evaluation of the European programme for industrial technologies and advanced materials (BRITEEURAM) shows that 63 % of partnerships continue between some of the partners, while 30 % continue with all partners originally involved.

III) Benefits concerning the socio-economic exploitation of results

Beyond the production of high quality scientific knowledge, the impact of the R&D programmes on industrial competitiveness will depend, amongst other things, on the capacity of researchers and industrialists themselves to transform knowledge into know-how and know-how into added value.

Acquiring or keeping competitive advantage implies a permanent adaptation to the needs of the market. It also requires an aptitude for integrating new technological opportunities and a rapid adaptation of industrial structures and the whole range of productive elements (a combination of factors of production, exploitation of R&D, marketing, etc.) in order to keep or increase market share in the face of competition.

The world economic environment is charging rapidly with the effect of globalisation of markets and company strategies as well as geopolitical changes. The economic scenarios which underlay the strategic ambitions of industries have to be constantly revised to remain valid.

- Better integration of research and application.

Commercial success or failure of a technological innovation requires the eventual bringing together of a whole number of factors. This is more likely obtained through the creation of partnerships and consortia. Indeed the fact of joining in the same project partners of different background from research to industrial demonstration, production and marketing is strongly increasing the probability of application of research results.

Experience has shown that the best economic results are obtained by consortia which include research laboratories, producers, suppliers and end users.

One factor in the success of the projects was an earlier contact or previous co-operation of the partners. In such cases, the communication and the co-operation were easier and allowed a better allocation of tasks.

Patenting behaviour is correlated with the efficiency of the co-operation and with the permanence of the partnership. This means that it is important that partners have no difficulties in the legal rights for exploitation of project results. Particular attention should be paid to partnership agreements concerning ownership of results and rights to exploit.

- Structure of collaboration.

It has also been shown that vertically integrated partnerships, where basic research institutes, producers, users and suppliers work together, are significantly more effective for all concerned parties. Considering the transfer of the achieved R&D results to their effective exploitation, two variables play a key role: the presence in the partnership of interested industrial end users and the will and capability of the partners to exploit the results. It is particularly important to involve in the consortium a "producer" partner capable of transforming a technical result into a commercial application. A statistical analysis divided projects into two groups, those with producer partners and those without. It was found that projects with producer partners:

- had higher capacity for exploitation;
- acquired more know-how;
- were better informed on exploiting their results;
- resulted in greater economic gains;
- can find applications on the market h a shorter time.

Furthermore when producers and users were both involved, the study found out that the economic effect of a given investment was in average 3 times higher than one of the projects where users and producers were not associated.

- Opening of market opportunities.

The establishment of international relations has had the consequence of opening more market opportunities. While collaboration between research institutes and industries is promoting the commercial utilisation of research, the inclusion of partners from different countries has geographically widened the marketing opportunities.

- Enhancement of spin-off through consortia.

The consortia created within the R&D programmes of the European Commission have also economic effects beyond the technological and scientific successes of the programme itself. They have the capacity to integrate heterogeneous partners and organisations with diverse knowledge and know-how and therefore are able to increase the occurrence of economic benefits. These benefits are produced by a complex process, including the dissemination of results, which can only be fully traced after the R&D programme has been completed and the major economic benefits have been worked through.

Studies have also proved that, even in cases where the main technical or scientific goals have not been achieved, there have been indirect economic effects which were often sufficient to justify the investment. The effort required for establishing international proposals has also proved a useful exercise. Indeed we found several examples where the partnerships created for submitting proposals went ahead even without having obtained a European financial support.

4. Conclusions

From the early preparation of R&D projects down to the exploitation of results an association of scientific and management skills is required. This means that training workshops on management, patenting and exploitation have to be regularly organised for scientists.

Furthermore the importance of partnerships shows that the inclusion of Russia in international R&D actions is likely to produce benefits of the type of those we have been able to measure in the European Union, provided sufficient attention is devoted to this aspect of project management.

R&D has to be from the beginning exploitation oriented. This requires the strategic interest of the project for the partners, their commitment and their exploitation capability to be monitored and checked during the projects. The identification of deficiencies should provoke quick reactions. An exploitation oriented management has to be ensured. Partners should be stimulated to identify markets for the results of the projects as early as possible and particular attention has to be paid to the vertical integration of different types of partners.

References:

1. *L.Bach, M.Ledoux. R.Magnaval, H.Pero.* The economic effects of public/private co-operative research.
2. *European Commission.* BRITE & EURAM evaluation study of finished projects: Synthesis report EUR 14541, 1992.
3. *R.Magnaval, L.Massimo and J.Removille.* Evaluation of co-operative research. Futures, Vol. 24, №. 7, Sept. 1992.
4. *J.Hagednorn and J.Schakenraad.* The economic effects of strategic partnerships and technology co-operation. EUR 1395O. 1991.
5. *L. Massimo.* The use of indicators in the R&D evaluation activity of the European Communities. Scientometrics, Vol. 21, №. 3 (1991). p.255–262.
6. *F. Narin, K. Stevens and E.S.Whitlow.* Scientific co-operation in. Europe and the citation of multinational co-authored papers. Scientometrics. Vol. 21, №. 3 (1991). p. 313–323.

POLISH EXPERIENCE WITH INTERNATIONAL MANAGEMENT TRAINING PROGRAMS

Tadeusz KOZEK
Task Force for Training and Human Resources
Cooperation Fund, Warsaw, Poland

1. Introduction

Management education has become an important part of the economic, social and political transformation taking place in Poland in the 90's. It has become a visible symbol of this transformation for two reasons. Firstly, due to the rapidly growing demand for management cadres, the number of management schools and management courses offered has increased dramatically since 1990. Secondly, this demand has created a market opportunity quickly grasped by entrepreneurs opening business schools, management institutes and centres. These institutions offer management education of various standards and levels and compete directly with the traditional education system.

Management education in Poland has a long-standing background and tradition. Before the Second World War, management education in Poland was already present and supported by the Institute of Scientific Organization. The leading figure in business education, Karol Adamiecki, established the influential monthly "Organizational Review" in 1926 to promote management theory and practices. After the Second World War, due to the communist regime, only limited attention was paid to the issues of managerial education, especially in the fields of marketing, finance and financial accounting. This does not mean that a system of business education did not exist. On the contrary, in the sixties and seventies three main sub-systems of business education were created: state, cooperative and academic.

The state system consisted of centres primarily training administrators for public administration and middle managers for industrial enterprises and institutions (e.g. research and development centres). This was built on the hub and spoke principle. In the process of training public administrators the role of hub was fulfilled by the powerful Institute of Administration and Management located in Warsaw – a research and training institution employing up to 300 researchers and lecturers. The institute developed programmes for public administration education that were later used by training schools in ministries (e.g. the Ministry of Payments and Social Services), provinces (voivodships) and districts. The training centres organized a broad range of courses: from regular executive training sessions to basic, skill-oriented programmes.

The process of training middle and top management in industry was organized in a similar fashion. The leading role was fulfilled by industrial centres for management

C.M.R. Verkoeyen et al. (eds.), Management Training in High-Tech and R&D, 115–125.

development. Each industry had its own centre that controlled, coordinated and supported the activities of local training centres.

The system of cooperatives that dominated the construction and food processing industries developed its own training system. This consisted of several training centres, usually well-equipped with libraries, lecture rooms, etc. These centres primarily trained middle managers of the cooperatives and also cadres of the Association of Cooperatives – a coordinating body supervising the operations of several cooperatives in the same industry.

An academic system of education was developed separately at universities and polytechnics. At the polytechnics, the model of the Industrial Management Institutes dominated, one that trained people in technical and production management. At the universities, the business school model, with a trend toward economics and law, dominated. The most influential and leading institution was the School of Management at the University of Warsaw. This was established as a joint venture between University of Warsaw and the Polish Academy of Sciences and in 1973 became a fully-fledged school of management at the University of Warsaw. It employed as many as 102 researchers – full, associate and assistant professors, in seven major departments: organization theory, national economy, company organization and management, public administration and law, quantitative management techniques, management systems, and marketing. Each department was divided into several sections specializing in particular topics (e.g. the organization theory department consisted of four sections: methods and techniques of management, organizational sociology and psychology, methods and techniques of work, and organizational cybernetics). During the seventies and eighties the School of Management pursued a rational development strategy, maintaining close ties with several US business schools and constantly training its staff abroad (mainly in the USA). Several of the professors from the School of Management taught in European and American business schools, importing skills, know-how and even the latest trends to the University education system in Poland. The range of courses offered by the school was very broad: from the basic five year Masters degree course in management to several general or tailored postgraduate courses for particular firms, with a limited number of students. **I want to stress the importance of this and other management schools or departments at the universities, because they became the main supply of trained lecturers for all business schools in the 90's.**

Apart from the three sub-systems described above, there were a number of other organizations and institutions deeply involved in managerial education. Probably the most important roles were fulfilled by national associations e.g. the Scientific Organization and Management Society (*TNOiK – Towarzystwo Naukowe Organizacji i Kierownictwa*), the Polish Economic Society (*PTE – Polskie Towarzystwo Ekonomiczne*), the Chief Technical Organization (*NOT – Naczelna Organizacja Techniczna*). The statutes of national associations made them active in areas of research, scientific information exchange and dissemination and management training on selected topics.

To summarize, in spite of the relative lack of interest and serious need for managers trained in modern methods and techniques, Poland had a substantial and well organized management training system. In effect there were several well-developed management education programs implemented, especially Masters programs in industrial or general management at the polytechnics and universities. Economists, public administration specialists and lawyers specializing in administrative law predominated among the staff of academic business schools, but a limited cadre of well educated management scholars, trained in western countries and often commuting back and forth, teaching in Polish, European and US business schools, was also available. **However we must stress that in spite of a massive educational effort, this had a limited effect for two reasons. First of all, such topics as finance or marketing were underdeveloped or entirely non-existent. There was no need or demand for such knowledge in a planned economy. Technical education and general management topics dominated. Secondly, most of the participants considered management training a routine and symbolic event of little practical value.** The eighties saw the start of the breakdown of the old structures of management training in Poland. This accelerated in the late eighties for economic, political and institutional reasons. The state sector that trained both public administrators and industrial managers collapsed. Ministries and enterprise associations were dissolved or merged and many training centres likewise merged. Even this process did not help substantially, since neither ministries nor enterprises could financially support their existence. Therefore many of them had to downsize and were later closed down. The public administration management training system was dissolved, a process begun with the decision to close down its hub. The Institute of Administration and Management was closed down by the government and its employees fired.[1]

Later on, in the 90's, a process was begun for the reconstruction of the system of public administration training, with the establishment of an elite National School of Public Administration.

The powerful and extensive cooperatives training system lost its economic basis in 1990 and 1991, when the political decision to break down major, national cooperatives was made. This was a sound decision, since most of the cooperatives were really massive. State-owned enterprises "dressed" in the cooperative legal form. However, financial problems, downsizing and the breaking down of the cooperative structures and the generally hostile climate toward the main cooperatives resulted in the almost total disappearance of their training structures.

National associations were put on the spot in the 1989 and 1990. As associations created or supported by the former, communist regimes, they were not favourites of the new power elite. However, they were not as easy or important targets as the industrial or cooperative structures. They began to adapt to the new environment and due to local initiatives, they partially lost their centralized character, becoming primarily a loose federation of local offices coordinated by elected central bodies.

The academic system (being politically neutral and institutionally shielded) was left intact, and slowly began to develop and gain importance with a growing understanding of the relevance of new western-type managerial training.

To summarize, 1989 saw the extinction commence of the old "dinosaurs" of management education. Massive structures of management training within industry and the cooperatives collapsed, were then downsized and dismantled. The painful

[1] In 1989 the Institute was considered by the new Mazowiecki government as a powerful institution controlled by an "old guard" of apparatchiks. Therefore, instead of efforts to adapt its role, the decision was made in 1990 to close it down and rebuild the system from scratch.

transformation of the public administration training system was begun with the closing down of its former hub and many spokes.

Other national institutions slowly began to adapt to the new conditions. Only the academic structures were left relatively unchanged but they too had to eventually change and adapt their courses. Being academic institutions, they quickly discovered the need but only very slowly began to take action, which opened a window of opportunity for private institutions and organizations that targeted managerial education as their field.

Since 1989 Poland has seen impressive development of private management education. Beginning in 1990, many schools were opened as fully-fledged schools or just centres, institutes etc. There were two main reasons for the rapid development of a management education "industry" in Poland. The first and simplest reason was a growing demand for such an education. The second was a liberal legal environment, especially in relation to postgraduate education. Two acts regulating educational issues: (1) the Act on the Educational System, dated September 7th 1990, and (2) the Act on Higher Education dated September 12th 1991, treat postgraduate education only marginally. In order to start educational activity, all that is needed is a permit from the local Education Board. The combination of demand and the lack of any legal constraints resulted in a proliferation of both private firms that called themselves business or management "schools", and private universities entitled to supply a 3-year bachelor-type education.

Foreign assistance has contributed significantly to starting up and sustaining the said development. Its size, however, was not so substantial as public opinion realised.

It is also difficult to assess its role in a unambiguous way, both because of its differentiation in terms of forms and scope, and its continuing evolution in the period of the last six years of system transformation. These issues will be dealt with in next chapters of this paper.

2. The scope and forms of foreign assistance in education and training

Foreign assistance is usually seen as a transfer of funds from one state to another. The transfer neither results from market operations nor carried through according to market principles [1]. The present article concentrates only on so-called non-reimbursable assistance in the form of dotations requiring no repayment.

Non-reimbursable assistance may be divided into three categories, according to its source:

- bilateral assistance offered by individual governments;
- multilateral assistance from international organisations or financial institutions;
- assistance from private sources.

Non-reimbursable assistance is granted in the following forms:

- financial subsidies;
- in kind (natural goods);
- technical assistance.

Poland has been receiving foreign assistance since the latter half of 1989. The process was initiated by a Paris meeting of the G 7 states in July 1989, where it was decided to entrust the European Commission with coordinating aid for the reforming Central European states with all the G 24[2] countries contributing to some extent. Education and training became a priority.

Technical assistance is the main form of bilateral assistance offered by the individual G 24 states. In practice, this means that assistance funds are granted to organisations inside the donor-country, which offer Poland not money, but counselling services and training schemes. Assistance in the form of technical equipment or training utilities is rarer.

The PHARE program[3] is the main source of multilateral assistance. Created in 1990, PHARE grants subsidies to finance individual programmes in accordance with conditions set by the European Commission and the recipient country's government. Program management is not centralised, and is carried out through a network set up inside the recipient country. Within this network, project funds may be distributed as agreed and in keeping with procedures set by the Commission.

The exact size of the bilateral assistance funds granted Poland is unknown. This, as already mentioned, is mainly because these funds do not arrive in Poland themselves but in the form of technical assistance or natural goods. During the early stages, donors were also occasionally reluctant to disclose information about the distribution of funds to individual assistance schemes. This in turn made it hard to estimate education and training funds precisely. According to estimates by the Task Force for Training and Human Resources, to-date bilateral assistance in education and training has not exceeded 200 million USD, of which about 120 million were used to train and prepare business staff. Bilateral assistance programs have become less of late. The only meaningful bilateral ventures 1995 were schemes offered by Germany, France, Britain, Sweden, Japan and the United States.

Between 1990 and 1995, PHARE spent the following for its education program:

- 135 million ECU for the TEMPUS[4] programme;
- 29 million ECU for other education programmes.

These amounts exclude expenses for training strands contained in non-educational programs like education restructuring schemes, support schemes for small and medium-sized enterprises, etc. With the exception of TEMPUS, the education programs financed by PHARE were only marginally concerned with training management staff.

Private assistance usually comes from private foundations as support for non-government organisations. The education expenses carried by these foundations are

---

[2] A group of highly-industrialized Western countries: the 12 EU members of the time, 6 EFTA (European Free Trade Agreement) states and Australia, Canada, Japan, New Zealand, Turkey and the United States.

[3] Poland and Hungary: Assistance in Restructuring Economies

[4] Trans-European Scheme for University Studies

unknown. The total amount spent by 7 selected foundations [5] between 1990 and 1994 were estimated [2] at 28.2 million USD.

What is the scope of foreign assistance? As an example let us take state expenses for education in 1993. All-in-all, these expenses came to about 6,444 million Zlotys – (approximately 3 billion USD)[6]. The education and training costs said above to be carried by bilateral and multilateral foreign assistance programs amounted to about 400 million USD over a 5-years period (80 million USD – about 2.7 % of the state budget costs – annually).

3. To-date experience

As shown, the expenses carried by foreign assistance programs are not considerable compared to state education costs, and particularly in view of the existing requirements. These limited funds, however, have an import greater than quantity alone. This is chiefly because they can be used to finance ventures otherwise certainly out of reach for the countries in question in light of their economic hardships and limited education spending. These projects concerned international cooperation, support for new schools and training centres and reforming and upgrading education programs.

The projects were successful not only thanks to the funds at their disposal or the forms of extended assistance. No less important – maybe even decisive – was the implementation system used. Typical here was a tri-lateral arrangement between organisations in the donor-country, organisation in the receiving country and the final recipients. The following four-step international cooperation scheme [3], providing for tri-lateral cooperation between foreign and local experts and the recipients may serve as a point of reference.

Step 1: The "old arms" model, based on the famous 17th-century trade triangle between Europe (outdated, ineffective weapons), Africa (slaves) and America (rum, cotton, sugar). Applied to education, this model merely outlines the existing programs disregarding the adaptation factor.

Step 2: The colonial, or oasis model. As in the colonies, outside experts manage development and the recipients carry no meaningful responsibility. Everything soon returns to its former state once the experts leave the country. Education measures consist in foreign expert training for local forces. Step 2.5 is a variant where an efficient experts does everything to adapt to local reality. Whereas this may be done by a single expert, it does not reflect on the system as a whole.

---

[5] Charles Stewart Foundation, Ford Foundation, Cooperating Dutch Foundations, Pew Charitable Trusts, Robert Bosch Foundation, Rockefeller Brothers Fund, Soros Foundations, Volkswagen Foundation.

[6] Figures according to the Small Statistical Yearbook, GUS 1995. The costs have been estimated in "new" Zlotys.

Step 3: The "academy" model: This model provides for assistance by local experts, who are, however, unacquainted with the conditions experienced by the final recipients. Adaptation possibilities here are limited, although much more extensive than in Step 2.

Step 4: The platform model. Joint or mutual development ventures implemented by experts on both sides possessing practical experience with regard to the recipient's environment. All sides may benefit from this form of cooperation. Recipients are schooled by local experts and lecturers.

The above models allow for the following conclusions regarding to to-date foreign assistance in training managerial staff:

1. The "old arms"" and colonial models certainly dominated foreign assistance during the first stage (1990–92). Both sides were responsible for this development. In Poland, the hopes connected with the initial period of transformation – seen today as naive – resulted in expectations of a "great leap forward" regarding human skills. Adapting personnel skills to new economic conditions was seen as a one-time, broad-scale move, based mainly on Western know-how purchased with the help of foreign assistance funds. The donors, on the other hand, frequently wanted to fulfill their assistance obligations as quickly as possible and spent their funds without properly researching local needs or conditions. Assistance programmes were also frequently left to commercial institutions, which, motivated by their own interests, offered "outdated weapons".

2. The "academy" model appeared more frequently during the second stage (1993–1994). This model is chiefly used in programs incorporating academic cooperation, although it can be applied in others. Participating local experts do not have sufficient experience regarding the recipient's environment – a typical example in Poland is leaving the training of managerial personnel to academic staff with no practical business experience.

The last of the mentioned models, the platform model, has become more widely used in recent years, its scope, however, is still quite small. Partnership-based cooperation of this kind is usually hard to realise under programs covered from foreign funds as the initial assumption here is that one side has more funds than the other. This concerns especially projects financed under bilateral assistance schemes. The situation is better under PHARE – particularly its TEMPUS strain. Under TEMPUS, the Polish side can really influence a project's aims, its implementation and fund distribution. Unfortunately, TEMPUS in its early stages did not provide for Polish institutions playing the part of "contractor" in disposing funds at hand. After this became possible, Polish partners in a large number of the approved JEPs[7] began to fulfill this function. This happened in 36 of 65 projects in the 1995/96 academic year and in 32 out of 91 projects in 1994/95[8]. On the other hand, organisations representing the economy proper, the actual direct or indirect recipients of most educational schemes, rarely participate in their realisation. Thus, according to the report [4], Polish industrial enterprises and economic institutions cooperated on only 26 out of the 75 "Management

[7] JEP – Joint European Project

[8] Based on information from the TEMPUS Poland office

and Economy" schemes realised through TEMPUS (about 35 percent). This cooperation mainly consisted of:

- organizing practical training in industry for foreign students;
- sending own employees to training and study courses organised under the projects.

This kind of cooperation does not generate sufficient interaction to effectively improve the practical aspects of higher school education programmes. More cooperation by the economy in ventures aimed at working out new, or upgrading old education programmes would change a lot in this respect.

4. Another important criterion in evaluating the successfulness of international assistance programs is their contribution to restructuring managerial training in the recipient country. To-date experience shows that:

- The growing numbers of education centres and programs in recent years were, to a certain extent, connected with foreign assistance, both bilateral and TEMPUS-sponsored. Examples here are four regional management centres created with the help of the British Know-How fund or post-graduate courses in French organised by the France-Poland Foundation (29 programmes, including 24 in management and associated fields [5]. TEMPUS funds have helped to set up twelve new education centres [4].
- Certain non-state schools and training centres benefited considerably from bilateral assistance schemes. This is particularly true of cases where foreign donors deliberately invested into complex projects in the non-state sector (e.g. the Know How Fund) or when non-state schools or training centres knew to profit from limited forms of bilateral foreign assistance (e.g. lecturers) and invested thus-saved funds into their own infrastructure. This kind of long-term planning was used more frequently by non-state education centres than state-owned schools, which tended towards hasty exploitation of foreign assistance aid.

Characteristic for foreign assistance is its small influence on personnel training in enterprises. This is connected to a broader phenomenon of enterprises withdrawing from training schemes for their crews, particularly during the early transformation period.

4. The role of training in polish enterprises after six years of system transformation

As mentioned earlier, the first phase of the transformation was characterized by a marked reduction in staff training in enterprises. The following factors appear to underlie this state of affairs:

- the weakening position of services responsible for staff development and their inability to function under the new economic conditions; as a result, in many enterprises there are now no staff who can deal competently with human resource development;

- financial constraints: due to their poor economic condition, a large number of enterprises are unable to allocate their own resources for staff training; neither can they receive support from the government agencies which they could count on under the previous system;
- the reduction in enterprises' own training capacity as well as the fact that the former sector-based system of staff training for the needs of the economy has at least partially broken down. Those elements of the previous training system which are still functioning do not often catch up with the economic transformations taking place.

These observations apply mostly to state-owned enterprises. Privatized enterprises are usually more capable of identifying their training needs and meeting them through participation in various privatization projects supported by the State and, in many cases, also through foreign assistance. The newly established private enterprises do not normally consider training as an economically advantageous instrument of their human resources policy; a more advantageous strategy for them is to employ new staff with the required skills rather than deploy lengthy and costly training of those who are already employed. A different approach can be observed in joint ventures involving western capital and branches of western enterprises, which import from the West modern instruments of policy for human resources development. Especially in the case of the latter, this means training of local staff in western in-house training centres and practical placements in the enterprise headquarters.

Putting aside the problems strictly related to the system transformation, it is clear that the processes of adaptation of enterprises to new conditions, characteristic of the more mature market economies [6] have already been to some extent developed in Poland; these result from the fact that:

- the processes of concentration, merger and co-participation of various entities in economic activities continuously change the conditions for competition; and this is reflected in the way that some enterprises perceive their shrinking or rapidly changing markets;
- technological and organisational changes are often so great and sudden that they turn out to be profound transformations, which force a large number of enterprises to change their main business activity policies.

In this context it may be observed that in Poland too, the sectors of the economy and the companies which are less exposed to the effects of competition have made much more limited progress than others with regard to forecasting the consequences of the inadequacy of the number and skills of staff under new conditions. The situation in most enterprises, however, is still far from standards seen in highly-developed economies. A report [7] based on research carried out in Polish enterprises provides some interesting information on the subject. Over a half of the polled managerial staff admitted that their enterprises did not designate any funds at all or only symbolic sums for managerial training schemes. This situation is especially true for small cooperatives and private companies. Big state-owned and privatized enterprises have the largest training funds. These, however, are still very small compared to the amounts put aside for training in the EU (500 to 1500 ECU per employee). Over one-fourth of the polled managerial staff had not taken part in any kind of training course within the last two years and over 31 percent attended one or two courses. The training offer consists mostly of short seminars as well as academic-oriented conferences and seminars, clearly a throwback on traditional contacts with higher schools and research institutions developed under the former system.

5. Final remarks

Poland, just like other countries of Central and Eastern Europe, was faced with the necessity of conducting a substantial reconstruction of its training system. This process has proven to be significantly more complicated and long-lasting than was initially assumed. The interdependence of this process with the development of the socio-economic structures, educational and training needs have not been treated as priorities thus far. Nevertheless, significant progress has taken place in this area, above all thanks to the introduction of a new legal framework which has made the education and training system more democratic, more decentralized, and more liberal. The grass-roots and the market mechanisms which has been set into motion have allowed progress to be attained in this area. Foreign assistance has also played an important role in this process. However if Western taxpayers want to be sure that their money is spent wisely and efficiently, the West should continue to listen to what our needs and preferences are, as well as farther modify the mechanisms and channels of transferring assistance. In particular, more attention should be given to the following factors:

- avoiding allocation resources to projects which can be implemented with national funding and capacities;
- the broader participation of national experts in preparing and executing programmes with a view to improve suitability of the latter to the recipients needs and local conditions;
- flexibility of the forms of assistance, including direct investment in teaching infrastructure;
- flexibility of procedures;
- more careful selection of foreign partners, frequent monitoring of programmes in progress and an independent evaluation of their final effectiveness;
- developing links between education and training systems on the one hand and labour market and economy, on the other.

References

1. Barre R., Luers W. H. et. al.: *Final report of the IEWS Task Force on Western Assistance to Transition in the Czech and Slovak Federal Republic, Hungary and Poland: Moving Beyond Assistance,* Institute for East-West Studies, New York, 1992.

2. Ners K. J., Buxell I. T. et. al.: *Assistance to Transition Survey, 1995,* IEWS Policy Education Centre on Assistance to Transition, Warsaw, 1995.

3. Freltoft T., Kay J.: *The Platform Philosophy: A Book about Synergy in Transnational Development,* AMU – International of Copenhagen, 1993.

4. Wirkus M.: *Report on JEP's in management and economy,* study commissioned by TEMPUS Poland Office, Warsaw, 1995.

5. *Francophone Masters in Poland 1990–1995,* 3rd edition, Foundationa France-Pologne and French Embassy in Poland.

6. Thierry, D.: *Restructurations et reconversion. Concepts et methods.* Editions l'Harmattan, Paris, 1995.

7. Karpowicz E., Szaban J., Wawrzyniak B., *Polish Manager in 90's*, Research Report, Warsaw, 1996.

TECHNOLOGY AND BUSINESS AUDITING OF SMALL AND MEDIUM SIZED ENTERPRISES

Wilfred Herbert WHITLOW
NIMTECH, United Kingdom

One of the themes running through this conference, as in most conferences on this subject, is the difficulties faced by Small and Medium Sized Enterprises, SMEs, in the start-up phase. These difficulties are very similar regardless of which country the SME is in – Russian Federation, Turkey, Thailand, United Kingdom etc. The organisation I represent, NIMTECH, was set up primarily to help SMEs in N.W. England during this critical phase, and of course when they are established in the need arises.

NIMTECH is located in St. Helens in N.W. England on the borders of Merseyside (Liverpool) and Manchester. In the mid-1980's when NIMTECH was set-up, many of the traditional industries in the area were downsizing or even closing, and the surplus staff were setting up their own companies, i.e. becoming SMEs, It was a kind of knee-jerk reaction, and many of the new managers of the SMEs had not thought through the implications and consequences of running their own companies. So the failure rate of these SMEs was high.

NIMTECH started in 1986 with a staff of 5 people, mostly from the management teams of major organisations in the area. We now have a permanent staff of 28 and we have 25 consultants who can be brought in when the necessary expertise is not available in-house. We are no longer confined to the N.W. Region of England. We operate throughout the country and we recently formed on international division with offices in France, Germany, Spain, Portugal, Italy, Sweden, Hungary, Brazil, India, and we will shortly be opening offices in Malaysia and in the USA. The number of companies that we assist in one way or another is about 2000, not all SMEs. We frequently help the larger companies and organisations to obtain funding from the various European Commission Programmes.

Nevertheless, the majority of organisations we help are SME's and our main aim is to help them with technology transfer. However, before we assist any SME (or larger organisation) whether it is just starting out or is already operating, we conduct a technology and business audit to establish whether they have the right type of infrastructure to develop and expand. How we do this is the main subject of my talk today.

For a very small SME, the audit might take half an hour, but for an SME of 40 or so people, it usually takes a full day. The time required to analyse the data varies

C.M.R. Verkoeyen et al. (eds.), Management Training in High-Tech and R&D, 127–131.

Overall objective

To establish whether an SME has an appropriate infrastructure for development and expansion

Fig. 1

Methodology

1. **Pre-visit questionnaire**
2. **Company visit**
3. **Analysis of data**
   - **SWOT**
   - **Calculation of Technology and Business Profile**
4. **Conclusions and recommendations**

Fig. 2

depending on the depth required. We have done over 500 audite world-wide, some in the Russian Federation. Of course, each audit has to be prepared specially for each SME, or group of SMEs. The questions put to an engineering SME would be different to those put to a biotechnology SME.

Our methodology is simple. It has four stages, The first thing we do is to ask the company to complete a short questionnaire – 2 sides of A4, 10 questions, asking for basic information about the SME; name, address, telephone number etc., description nature of product/service, profits (if they are already operating), market, etc.

Surprisingly, some SMEs have difficulty in completing the form. It can be argued that the questionnaire is a first test of the SME. We then visit the SME (I will come back to the issues we discuss in a moment). Back at the office, we analyse the information we collected during the visit, perform a SWOT analysis, prepare an Technology and Business Profile (again, I will come back to this in a moment) and then draw our conclusions and recommendations. We are always brutally honest with the SME; maybe their infrastructure is inadequate in some way, their business is a non-starter, they do not have a management team with the appropriate technical or business experience, etc.

All the questions we ask during the company visit are simple and logical, but the answers when put together give a very clear indication of the SME's profile, its strengths and weaknesses.

Methodology: Company visit

- **Company image**
- **People**
- **Products**
- **Manufacturing**
- **Quality**
- **Resources**
- **Finance**
- **Marketing and Sales**
- **Training**

Fig. 3

These are the areas we talk about during the visit. Within each area, there are several questions, each of which we assign a rating to. I am not going to go through all the questions in each of the 9 areas, but I will just concentrate on three:

Company Image (A)
Quality (E)
Training (I)

Methodology: Analysis of data

- **SWOT**
- **Technology and Business Profile**

Fig. 4

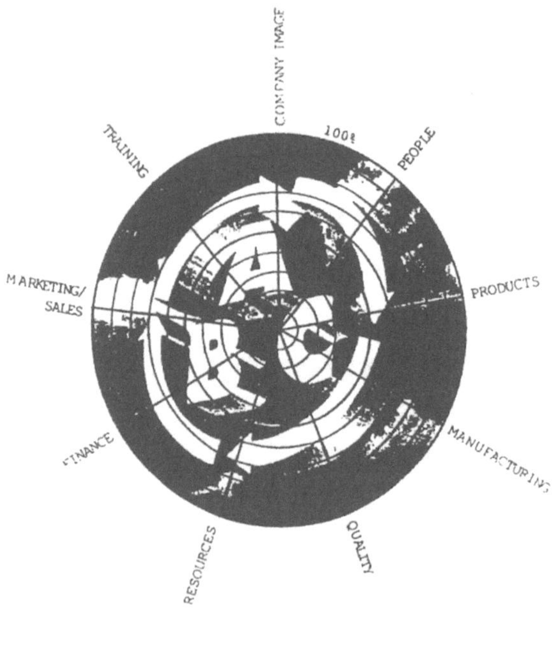

Technology and Business Profile

Fig. 5

The next step (as I mentioned earlier) is to do a SWOT analysis of the company and to calculate its Technology and Business Profile.

This is blank Business Profile. Each spoke corresponds to one of the areas discussed during the company visit and our calculations give a rating for each area. The points on the spoke are joined up to give an “at a glance” profile of the company. The profile for an ideal company would be a full circle connecting the 100 % points. No companies that I have talked to have such a profile.

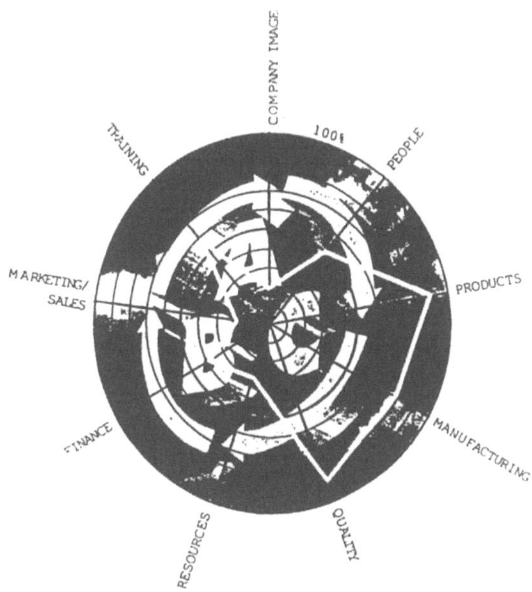

XYZ Autoclave Manufacturing Co.

Fig.6

The XYZ Autoclave Manufacturing Company is purely hypothetical but it is typical of many SMEs in the UK, good manufacturing facilities and a quality product, but a poor company image, and a management team weak on financial matters, and with inadequate marketing and sales skills and activities.

So, it is clear what types of training are required.

COOPERATION OF RUSSIAN AND FOREIGN SPECIALISTS IN HIGH-TECH MANAGEMENT TRAINING AND CONSULTING

Andrei FURSENKO
SPb. Regional Foundation for Science & Technological Development,
Albina NIKKONEN, Oleg KABANOV
St.Petersburg High-Tech Management Training Center

The critical situation existing today in small and medium business, especially in high-tech-oriented firms, has been caused by both external (objective) difficulties and internal (subjective) problems related to the operational organization of the firms themselves. The latter include some lack or occasionally total absence of qualified management staff.

This problem is faced by practically all potential investors and partners of small and medium high-tech enterprises (HT SME). It is quite natural that St. Petersburg Regional Foundation for Scientific and Technological Development (RF STD) established in 1992 as a non-profit organization for assistance in the preservation and development of the R&D potential of St. Petersburg and Leningrad region under transition to market economy had to come across this absence of appropriate management in high-tech sphere, in general, and in innovation sphere, in particular.

The experience accumulated by RF STD in its work with applicants shows that to successfully implement high-tech projects managers need additional skills and practical assistance in key matters of marketing, finance activities of their companies, strategic management proceeding from specific challenges of high-tech products development and production under severe internal and external competition.

Today it has become clear that not a single original idea, not even the one having a competently developed business-plan, will work if there are no specialists capable of managing the commercial aspects of R&D programs and bringing them to concrete practical results.

One of approaches to such specialists training has been undertaken by St. Petersburg Center of High-Tech Management Training (Training-Center) established in 1994. RF STD is among the founders of the Training-Center.

Setting up of the Training-Center was caused by basic changes taking place now in high-tech sphere, in particular, in defence industry where a great number of highly qualified researchers leave large enterprises and research institutes and start the implementation of their own projects within the framework of newly established small and medium firms.

And very often it is not the lack of actuality, significance or profitability that becomes the main obstacle in the successful implementation of such projects, but the form of their presentation on both international and domestic markets, the methods of their promotion, the understanding of their commercial attractiveness by their managers. Therefore, at present in risk estimation making a part of application analysis

C.M.R. Verkoeyen et al. (eds.), Management Training in High-Tech and R&D, 133–137.

the main attention is paid to staff competence and ability to withstand all difficulties arising at all stages of an enterprise development.

The success of a project, a very often of a firm as a whole, because as a rule at the initial stage the firm is able to develop just one activity, strongly depends on a manager's personality, his professional fitness. Today a manager must be, first and foremost, a manager, and not a scientist in a firm. Though, certainly one thing does not exclude the other.

Their specialists training makes today the main, if not the only way to solve staff problems at HT SME, as labor market in this field remains rather limited.

The regular training, generally accepted in developed market economies, implies that young people graduate from universities, get business-education and then come to science or industry to acquire practical experience and special skills necessary for proper managers. This way of training can give results not sooner than in 5–7 years. It should be emphasized that during this time the major part of technological ideas and R&Ds which can make a basis for HT SME's operation will become hopelessly obsolete. So, a task was set to support researchers having gift for organizational activities, to provide them with special knowledge of practical application, to orient them to taking consultations in specialized (marketing, intermediary, legal) firms offering such services.

In 1995 the German government made a decision to finance the project "Training and practical on-the-job training of high-tech management" introduced by the Training-Center jointly with one of its founders – the German firm PSB.

The training program was developed by Russian and German specialists and intended for training all members of R&D staff. It was oriented to trainees' projects implementation, finding additional investments and strategic partners. The final result of such training and consulting was to be estimated by the market success of projects the managers of which would take the full course of training.

From the very start the program was drawn up proceeding from the needs and preferences of trainees themselves and most of their requirements to training arrangement were taken into account. On one hand, that increased the effectiveness of the training process, on the other hand – created a number of organizational difficulties.

First of all, these were problems connected with trainees themselves. They were caused by peculiarities typical of all of them:

1. Being managers, i.e. key persons in their projects, trainees cannot stay outside their main activities for taking general or even special courses for future – they have to make management decisions already today.

2. These managers are interested not just (and from their point of view – not so much either) in getting basic information, but in consultations on different aspects of their projects implementation.

3. All newly established small enterprises find themselves in strained financial circumstances and especially – high-tech enterprises where the period of return on investment or credit repayment makes not less than 2–3 years. In this connection they cannot afford training and consultations of qualified experts at their own expense.

The second block of problems is connected with the right choice of lecturers. These problems may be formulated as follows: absence of appropriate methodical training and, often, lack of special practical knowledge featured by possible lecturers and consultants go along with high payment requirement.

All that is equally true for both foreign and Russian lecturers and consultants. The experience of work with training staff proves that the best result is rather achieved by lecturers having practical experience of projects implementation, than by those having classic academic knowledge. Still here the role of foreign specialists in domestic management training should not be exaggerated, as it often happens due to inertia in public thinking. Though their participation in re-training process is quite important and has its strengths. This concerns the following directions:

- search for the ways to enter the international market, marketing research;
- preparing a projects for their introduction on the international market, as foreign specialists feel better what moments should be dwelled upon or emphasized in project materials for the work with foreign and international financial institutions;
- getting ready to the work on the domestic market which is just being created, but already sets the rules in keeping with the international market;
- sharing the experience of establishing production under market conditions, as Russian specialists have difficulties in getting over the planned relations system of the past.

It should be mentioned that without the participation of Russian coaches all these issues remain estranged from the Russian reality. Not a single foreign specialist, no matter how educated he is, can ideally orient a project to the way which will lead to success.

The experience of the Training-Center in management training conducted jointly by Russian and foreign lecturers has shown the differences in understanding of many aspects, in approaches to high-tech projects development and commercialization, in preparing project managers for the work under market conditions. Proceeding from the opinion poll carried out among the trainees we can point out the following drawbacks in the work of foreign lecturers and trainees' problems which stem thereof:

- the fact that foreign lecturers repeat well-known material just annoys people and is interpreted by them as lack of respect to Russian trainees;
- Russian project managers have to flexibly adjust to changing environment as a result of which in their development they pass ahead of foreign lecturers' ideas about their progress in training and understanding of market economy;
- in teaching advertising techniques foreign lecturers fail to take into account Russian specifics (the said specifics is connected, first of all, with the fact that the mentality and reality perception of Russian people differs significantly from the reaction and behavior of people in foreign countries);
- inadequate attention is paid to the explanations of terms and definitions used, i.e. to reducing different notions to a common denominator.

In this situation foreign specialists act rather as detached onlookers, than active assistants of managers in their projects implementation. Even lecturers from Eastern Germany where structural transformations are most similar to ours, occasionally fail to adequately estimate the current situation. This is most likely to be connected with the fact that in GFR transition to market economy, first and foremost, was considered as a process of social adaptation with corresponding resources and approaches. In this country transition to market is carried out by the most strict scheme oriented not to the rescue of those weak, but to the support of real market and strong teams capable of further independent development.

Therefore, by our opinion, the most acceptable variant of arranging the training process is in the form of joint work of Russian and foreign specialists illustrating their explanations by concrete examples of the situation in which trainees' projects are now.

In this connection coaches should have fairly good practical knowledge acquired by them in going through a hard school of small firm managers' experience, and also be involved in practical work as consultants – financiers, lawyers, marketing specialists, etc.

The same strict requirements should be met by the methods and programs of courses offered for the training process.

We see the training process as ensuring the support and development of the project from its start-up to implementation. Certainly, it would be impossible without the participation of both federal and regional governmental authorities and international organizations. A good manager cannot be trained and, consequently, a good project cannot be obtained without the utilization of the whole world experience of such work, without setting close cooperation contacts between Russian and foreign enterprises.

It has already become clear that without imparting international character to projects, without establishing close contacts with Western partners firms will not be able to reach the world level in R&D introduction, production and to create competitive products.

Therefore, the methods of training and consulting were based on the approach which allowed us not to be confined to just one partner or to lecturers from just one country.

Along with the execution of the basic Russian-German program we were working hard to attract other foreign partners. As a result of these activities the Finnish Government has made a decision on financing the project "Industrial Property Protection" introduced together with the Chamber of Registration and Patents and the British Council has provided support for the work we carry out jointly with the Ulster University.

The "guest" seminar "Intellectual Property Rights in Innovative Activities" conducted in Finland by the Training-Center jointly with the Finnish firm "IPEC" has provided first results. 25 representatives of Russian high-tech firms took part in the seminar. Along with the consideration of theoretical issues serious attention was paid at the seminar to practical work in the search for mutual interests and establishing cooperation. And now at least 5 projects are at the stage of signing agreements. At a return meeting in St. Petersburg interim results will be summarized and the experience of establishing cooperation between the firms of two countries in innovative sphere will be generalized. It is suggested to use the same scheme for conducting seminars within the framework of establishing cooperation with the firms of Great Britain and Belgium.

In summing up the activities of the Training-Center during more than a year the following moments should be emphasized:

1. The strengths of foreign partner were successfully used in the formation and training of a pilot group of trainees constituted of project managers with whom RF STD

and other friendly organizations worked. So, financial management and marketing were taught by German lecturers. A seminar on certification and standards was conducted jointly with specialists from Ulster. And the problems of intellectual property protection were discussed with the participation of Irish and Finnish specialists.

2. The role of Russia experts and lecturers both in training and in consultations was extremely important. Thus, the statistics of lessons attendance and trainees' opinion obtained via questionnaires show that the seminars conducted with the participation of Russian lecturers were most popular.

3. Seminars held jointly by Russian and foreign lecturers prove to be most effective.

4. And finally, our students are active in setting close contacts and starting cooperation with each other, which, certainly, provides a favorable effect on the psychological climate in the group and contributes to the development of the firms.

QUALITY CONTROL OF TRAINING: THE ROLE OF LECTURERS AND STUDENTS

Sveta KOSHARNAIA
TNP Management, Belgium

1. Introduction

During a transition period, the Eastern-European countries come across numerous problems of both economic and social character. The radical changes which have recently occurred have influenced the economic relations between enterprises, enterprises and the state, enterprises and customers. New forms of business activities have developed. One of them, small and middle enterprises, has appeared in many spheres, including scientific. Some of such enterprises are a success, as can be seen from the reports delivered at the seminar.

However, most of enterprises are having hard times: the demand for their products is going down (as more high-quality and less expensive goods can be brought from abroad), they do not have enough money to buy raw materials, no funds for reorganization, innovation, their productivity is low and at the same time expenses are high. The managers of enterprises are, as a rule, the same people who used to be at the head during the planned economy times. Of course, there are a lot of good managers among them who mastered the new economic language very quickly, but for the most part, these people worked practically all their life in the conditions of planned economy and their assimilation to the new environment is not very easy.

It is so often that a western businessman is surprised when he has to wait for an answer to his question about the price of the goods produced by an enterprise during several weeks or never receives it. As a result, one has an impression that enterprises are not interested in selling their produce at all.

To turn existing enterprises into flexible, financially healthy and producing structures which could be competitive on the international market, the managers of enterprises should be the first to master the new methods of work and should become "conductors" of a new production culture, new ideas.

The growing demand for highly-qualified managers gave rise to a great number of business-schools and training-centers. Institutes set up new faculties, seminars become a common and regular phenomenon. A considerable part of the financial support goes for the training of managers by western specialists and lecturers under numerous training programs financed by international organizations such as the European Commission, the World Bank, EBRD as well as under bilateral programs of various countries.

All these measures are directed at the training of a group of managers who could efficiently manage the work of enterprises in the new conditions and thus provide the successful reformation of the economy.

C.M.R. Verkoeyen et al. (eds.), Management Training in High-Tech and R&D, 139–144.

2. Particularities of small and middle enterprises

Small and middle enterprises differ from large ones not only in their size. Considerable differences can be found in their internal organization, in the way the business is carried out, in the market approach, innovations, personnel management, etc.

Let us enumerate some of their characteristics:

- small and middle enterprises are less bureaucratic;
- their organization is more flexible, without strictly determined functions and positions; not too many formalities;
- the exchange of information takes less time; fewer hierarchic levels; the manager meets with employees more often;
- the manager can be much faster in reacting to a situation and taking measures; the decision results are very fast to be seen;
- the quality of work is higher; a higher responsibility and a better control; one can immediately see who is responsible and for what;
- the manager of an enterprise has a clearer idea of what is going on at his enterprise;
- the personnel are not numerous; often one person possesses all the knowledge in one field; if such a specialist quits, this affects the work of a small enterprise much stronger than of a big one;
- more opportunities for creative activities, a faster realization of new ideas;
- the manager takes more risks;
- the manager pays more attention to the operative management than to the strategic management;
- financial means of the enterprise are insufficient to conduct long-term research and development without additional financing.

3. Personnel training: Small and middle enterprises

Personnel and related problems started to attract more attention during the last years. The personnel management covers not only the administrative work related with hiring, salary payment, but also with the development of the payment system, personnel search, estimation of the work and training of the personnel, the establishment of the personnel development policy.

For a small enterprise, personnel are as important as for a big one, may be even more important, since, having fewer employees, a small enterprise is in a greater dependence on them. Small and middle enterprises differ from big ones in personnel management and training, which is defined by their specifics.

The training of personnel, including managers, is of highest efficiency when it is related with production. Big enterprises have an opportunity to organize the training on the spot. They have the necessary educational infrastructure, their own teachers. The education program takes into account the specifics and needs of the enterprises. For managers, special programs are developed.

Small and middle enterprises do not have such training-centers, they have no teachers, in some cases even no personnel managers. That is why, they cannot develop educational programs or prepare educational materials by themselves, so they have to send their personnel to institutes, special companies and organizations involved in training. This entails a lot of expenses, which creates a certain problem for a small enterprise. Often they have to put up with the terms of standard courses which do not meet, to a full extent, the specific requirements of this or that enterprise.

In general, small enterprises pay less attention to the long-term planning or the strategy development. Most of time they are busy with the solution of current problems. In the matters of personnel training, small enterprises also proceed from the needs of the current moment, rather than from the strategy of the enterprises.

For a small enterprise, it is typical that both the managers and employees are wholly busy with the current work and it is often that they cannot be replaced. That is why they have to restrict themselves to a short training course or training right on the spot. Small and middle enterprises do not have big personnel management departments. The managers of small enterprises are aware of the importance of the right personnel policy for their enterprises and often take all the responsibilities for the solution of matters related with the personnel.

According to the data of questioning of the managers of Belgian small and middle enterprises [3], 78 % are convinced that managers themselves are to be responsible for the personnel. This was the situation in 50 % cases: only one person was responsible for the personnel – the manager of the enterprise.

So, the manager of an enterprise often takes this responsibility, but having in the first place the current management of his enterprise to look after, he has no time to properly handle the problem of personnel as this could have been done by personnel managers of big enterprises. According to the data of the same questioning [3], 25 % of managers find only one hour per week of their time to settle the matters related with personnel. Of course, it is not enough to constantly keep up with all the opportunities of the manforce market. This results in the lack of information and the difficulty to choose the right training courses.

Managers realize well that the training of personnel has an important role in the further development of an enterprise, but it is not always that they are able to find the best approach to this problem.

Due to above reasons, small and middle enterprises prefer to have their personnel trained outside their enterprises, i.e. at enterprises, in institutes and organizations specializing in personnel training. Among questionned [3], 65 % state that their employees are trained outside their enterprises.

Therefore, taking into account the specifics of small and middle enterprises, one can list the following important points to be considered in personnel training:

- personnel are often trained outside the enterprise;
- preference is often given to short-term courses or to training right on the spot; the employees and managers of small enterprises have to choose this way, since they do not have to stop working and can do their daily work following the advice and having the support of the specialists;
- personnel training is directed at the solution of concrete problems and must be used in practice at once.

4. On the quality of training and measures of its improvement

The choice of all kinds of courses grows wider and wider. How can one find his way in the ocean of seminars and courses? How can one choose the training course which is the best to meet the requirements of one's enterprise and is the most efficient? Shall some general program be developed to suit various types of enterprises or shall the needs of one, concrete enterprise be the basis of a training program? These questions concern both managers and personnel training specialists.

Western managers as well as their Russian colleagues might at some stage find themselves in such a situation when on having finished some courses, they did not have an impression that they had not wasted their time and had acquired some valuable knowledge. Such situations can be avoided if to pay more attention to the quality of both training programs and the training process.

According to the definition given in the international standard ISO–8402, "quality" covers all the features and characteristics of a product or a service which can satisfy all current or anticipated needs and desires. In other words, products and services are to meet the requirements of the customer they are intended for.

As for the personnel training, one should outline the following requirements to quality:

- knowledge and skills must find their application in practice and meet concrete requirements of an enterprise; managers of newly-created enterprises do not have extra time to spend on the courses which have very little in common with managers' concrete production activities;
- the subject must be covered in detail;
- the didactic material should be of high quality;
- the teacher should be capable to impart his knowledge to students.

Which measures can be taken to improve the quality?

Several points are important here. First, the didactic material should be improved with the use of a certain set of quality standards; quality control measures are to be taken, too. They can be established in the quality plan. Second, the quality of the training process itself should be improved. It is significant that both teachers and students should play an important role in the process of the training quality estimation and improvement.

Let us suggest as to what can be done to improve the quality before the training starts, during the training and thereafter.

Before the training:

To develop a personnel training program meeting the realistic needs of small and middle enterprises, it is necessary:

- to be in the constant know of what is going on at the enterprise. The needs of the enterprise related with personnel training should be analyzed to develop a program which would be the best to satisfy these needs. The managers of the enterprise can help teachers to specify these needs;

- to keep up with the new tendencies in personnel training and to choose the most prospective. This is a task of organizations and enterprises involved in personnel training. They know the training market much better and have respective specialists.

During the training:

During the training, one should pay a special attention to the establishment of the constant contact between the teacher and his students. The teacher should stimulate the activity of his audience by questions and drillings. He should try to avoid such a situation when one is doing the talking and all the rest are just listening.

After the training:

With the help of a special questionnaire, it is possible to find out, how the students estimate the training course. The questions may be as follows:

- Was the subject covered in full detail?
- How much does the illustrative material correspond to the subject of training and whether this material is good enough?
- What is the quality of the didactic material?
- Which important subjects (in students' opinion) were not covered?
- Which parts of the course were of greatest interest, which of little interest, and why?
- Was there any possibility to ask questions and how did the lecturer answer them?
- Is there any application of the obtained knowledge at their enterprise?
- Which training elements cannot be used in practice at once?
- How do the students like the rooms where the courses are held, the equipment, the illumination?
- What do they think of the length of the training period?
- Etc.

The analysis of this information permits detection of the weak points of a training program and identification of measures for its improvement. After the period of training, the latter can be considered efficient only if the knowledge obtained during this period is used in practice. That is why the role of the teacher cannot be limited by the walls of the class-room. The help of experts to make use of the obtained knowledge right at the enterprise can be considered as another element of training. On the other hand, this can help teachers to assess the quality of training and its application in practice.

5. Conclusion

Due to their specifics, small and middle enterprises, to have their personnel trained, are often forced to resort to organizations specializing in training. Both parties are interested in the appropriate quality of the provided training, so both personnel training specialists and student-managers should play a more active role in the quality control. Their cooperation will lead to the creation of such kinds of training which are of primary necessity for enterprises during the transition period and which will be more efficient than the programs borrowed from the west.

References

1. *J.J. Boonstra, M.I. Demenint, H.O. Steensma* (Red.) (1989). Organiseren en veranderen in een dynammisch world: begeleiden van veranderingprocessen binnen organisatie.
2. *A.J. Cozijsen, W.J. Vrakking (1986).* Handbook voor strategisch innovaren: een internationale nalans, Deventer: Kluwer/NIVE.
3. *R. Dencels e.a.* (1991), Mensen in KMO's, Konig Beudewi jntichting, KMO–Stidiecentrum, K.U. Brussel, Roularta Books n.v., Zellik.
4. *J.P.J.M. Kerssemakers* (1992). Sociale vaardigheden voor managers & leidinggeven/Investeren In: Communicatie, inzet personeel, struringskunde, Nelissen R.V. Baarnn.
5. *Imai, Masaaki* (1990). Kaizen: de sleutel van succesvalle cocurrentie, Daventer: Kluwer.
6. *H.B. van Terwisga, E. vazn Slujs* (1990). Opleiden voor de toekomei: onderdeel van berdrijfsbeleid, Samson Alphen aan den Rijn, Deurne.
7. *J.G. Wissema, L. Euser* (1988). Samenwerkingbi j technologische vernieuwing: de nieuwe di mensie van het management van innovatie, Kluwer Bedri jfswetenschappen, NEHEM.

EXPERIENCE OF LASER ASSOCIATION IN TECHNICAL SPECIALISTS TRAINING IN MODERN MANAGEMENT FOR HIGH-TECH INDUSTRIES

Yelena SLOMTCHINSKAIA
Laser Association, Moscow

Laser Association (LAS) was established in 1990 as a voluntary association of researchers, manufacturers and users of laser equipment. At present it unites 201 collective members and more than 400 individuals. LAS is an international R&D organization the main objectives of which are as follows: creating a single informational R&D environment, arranging and ensuring support for joint R&D work and cooperation in laser equipment production, developing international scientific cooperation.

Organizations of various types enter Laser Association: research institutes, institutes of the Academy of Sciences, joint stock companies, small ventures, clinics, etc. The membership structure of LAS is given in Table 1.

Table 1

Laser Association Structure

Institutes and design bureaus of the Academy of Sciences	8.3 %
Higher educational institutions	6 %
Research institutes and design bureaus of various industries	32.7 %
Industrial enterprises	8.8 %
Small ventures	40 %
Clinics	3 %
Other	1.2 %

It should be mentioned that today LAS unites almost all academic and branch institutes, as well as half of higher educational institutions with their laser departments dealing in the development of laser equipment, about one-third of small ventures manufacturing laser devices and about 10 % of clinics using medical laser equipment.

LAS has a number of regional centers in the following cities and towns: Bishkek, Vilnus, Vladivostok, Kazan, Minsk, Nizhni Novgorod, Nikolajev, Novosibirsk, St.Petersburg, Samara and Tashkent.

C.M.R. Verkoeyen et al. (eds.), Management Training in High-Tech and R&D, 145–150.

The main directions in LAS activities are as follows:

- Informational support of work in laser development and application;
- International cooperation;
- Assistance to its members in search for business partners and joining national R&D programs of CIS states;
- Training R&D specialists in the management of R&D, manufacturing and sales of high-tech products.

LAS has created a unique computer data base on laser equipment development and application throughout the CIS states on the basis of which 6 specialized catalogs were published:

"Laser Equipment Produced by Laser Association Members";
"Laser Art Treatment of Materials";
"Laser Measurements and Diagnostics";
"Lasers in Medicine";
"Laser Technological Complexes, Equipment and Technologies for Materials Treatment";
"Laser Sources of Radiation".

We published five editions of the reference book "Who is Who in LAS" in Russian. Two editions were published in German, more than thousand copies have been distributed in Germany. The reference book "Leading Laser Centers of CIS" was published in 1993 in English. In 1994 we made a computer versions of this reference-book including 500 institutions of CIS working in laser equipment field.

In 1995 we drew up and published (in cooperation with our German colleagues) a bibliography of Russian publications on lasers made in 1991–93. Now a similar publication for the period of 1994–95 is being prepared. Each year 2–2.5 thousand articles are published in CIS and remain unknown to the majority of foreign specialists, as few of these articles are translated into English. Bibliographic publications of LAS help our foreign colleagues to adequately estimate the role of our scientists in the development of laser physics, quantum electronics, non-linear optics.

To contribute to establishing business contacts we have arranged 20 exhibitions and presentations of laser technologies. Eight of the said exhibitions were international ones. The collective stand of LAS at the International Exhibition-Fair in Munich (1993) became the first complex presentation of our laser technologies abroad. The exhibition-presentation "LIC Russia'94" held in Moscow with the participation of 13 countries was a big success. 125 projects for international cooperation in laser equipment development and application were presented at this exhibition. As a result 10 domestic firms have entered the international market.

LAS arranged the exhibition "Laser-95" in St. Petersburg which also contributed to the development of scientific and commercial cooperation.

In the field of international scientific cooperation LAS conducted six seminars jointly with its German colleagues: on high-power lasers (Orlovo – 1992, 24 presentations), on laser control and measuring systems (Raubichy – 1994, 25 presentations), on modern management training of technical specialists (Moscow –

1993, 1994, 1995, in total – about 30 presentations), on Russian-German cooperation in lasers (Drezden – 1994, 20 presentations).

Since 1992 we have arranged more than 40 visits of foreign specialists to Russia and Ukraine. Specialists from Germany, China, Singapore, USA often visit our institutes and factories. It is evident that such meetings between the representatives of our industry and foreign specialists are necessary, as earlier our opportunities of such contacts were rather limited.

LAS headquarters help LAS members to join international scientific programs. That implies consultations when applications are made and search for foreign partners via the dissemination of information on such programs as EUROLASER, INTAS, COPERNICUS. Since the time when LAS became the Russian party coordinator of the cooperation with Germany, we, jointly with our colleagues from VDI-TZ, have prepared recommendations on agreements for R&D cooperation. This activity has resulted in 40 joint R&D projects implemented by Russian and Ukrainian organizations jointly with German laser centers with obtaining funding from the Ministry of Science and Education of Germany (in 1993–95 our laser centers got support from Germany in the amount of 2 million Marks), 4 projects approved by ISTC, 10 projects offered to INTAS and EUROLASER.

Recently we have set good contacts with the Russian Foundation for Assistance in the Development of Small High-Tech Enterprises. In fact, we urged small enterprises to address the Foundation with their projects. The results of such projects implementation could be seen in the Foundation's exposition on the collective stand of LAS at the exhibition "Laser-95" in St. Petersburg.

However, in the process of developing international cooperation with both CIS and foreign countries the main LAS's objective has become that of training specialists for work in market economy. While the knowledge of such methods and skills of their application make the key point under the conditions of present Russia and other CIS countries, our engineers and scientific specialists practically have not been taught any of them.

Since 1992, within the framework of the Russian-German project "Training and Re-Training of CIS Laser Specialists in Modern Management and Marketing", LAS has successfully carried out joint work with the Institute of Planning and R&D Studies (IES, Hannover) and Eastern-European training and consulting block (BBO, Drezden) with the all-round support of the Federal Ministry of Science and Technology of Germany (BMBF, Bonn) and Technological Center of German Engineers Union (VDI-TZ, Dusseldorf). First the system of short-term (10-days) courses has been developed and then since 1994 regular and on-the-job two-months courses have started with further practical training of the best graduates in specialized firms of Germany.

At present three cycles of "long" courses have been finished. About 100 technical specialists from Russia, Bielorussia, Latvia and Ukraine have got training by our program. It is significant that the training process provides not just an educational effect (graduates are ready for professional cooperation with foreign partners, they start to understand how they should manage their own companies), but also helps to start new cooperation between Russian and German enterprises. As a result, in 1994–95 eight joint projects were started.

The main changes which took place in the LAS educational program during the last year are connected, first of all, with expanding the range of trainees: now these are not only laser specialists, but also specialists in biotechnology, telecommunications and space research, as well as in other high-tech fields.

The distinctive peculiarities of LAS courses are as follows:

1. Good knowledge of "its range" of trainees. All these specialists are distinguished by good technical education, high professionalism in special fields of science and technology, but poor knowledge of market mechanisms for their products promotion. Therefore, the training program is aimed at solving concrete problems faced by the trainees, and lecturers and experts who conduct the training are chosen most carefully (see Table 2).

2. Clear practical orientation of LAS courses. Along with complex studying of management and marketing problems, getting information on the economic law, taxation, customs law and intellectual property protection trainees independently develop business-plans for their products promotion. In the very beginning of the training trainees independently prepare proposals for cooperation with a German firm or an institute for their R&D's promotion. According to these proposals places for practical training are chosen.

3. Convenient form of training. Proceeding from tough schedules of present-day managers and impossibility of their long stay outside their businesses the courses combine regular and on-the-job training (2 weeks of regular training, 7 weeks – on-the-job training, 1 week – regular training). Trainees get a big set of specially designed text-books for independent studies during the on-the-job stage of training. Foreign language is also mastered by trainees independently.

4. Serious attention to teaching trainees the skill of introducing themselves to both "domestic" and foreign partners, i.e. the skill of advertising their opportunities and reliability guarantees. General culture and education of people can be of great help here, while their absence may bring to naught newly set contacts at closer acquaintance of partners. Therefore, keeping to the business ethics rules is necessary for modern businessmen.

LAS experience in arranging training courses was estimated positively and a new Russian-German project "High-Tech Industries Management for the Specialists of Small and Medium Defence Enterprises" has been approved for 1996–1998.

Table 2

Training Program

Program Item	Brief contents of topics
	First regular training phase (2 weeks)
Management	Modern problems of management theory and practice
	International management-business cooperation with the European community
	Social and psychological testing
	Business ethics and business communication. Practice of business negotiations
Marketing	Marketing basics and stimulation of high-tech products sales
Legal regulation in economy	Basics of Russian and CIS taxation systems
	Basics of Russian and CIS economic law
	Russian customs rules for high-tech products import-export
Finance management	Finance management in market economy
	Enterprise and entrepreneur in market economy
R&D management	Patenting peculiarities in Russia
	Business-plan development
	R&D management peculiarities and investment aspects
	Intellectual property
High-tech market	Laser market. Structure and evolution
Foreign language	Testing and communication practice (English and German)
	On-the-job training phase (7 weeks)
Management	Studying training and methodical literature
	Analyzing management situations
Marketing	Studying training and methodical literature
	Analyzing management situations
Finance management	Studying training and methodical literature
	Solving tasks
R&D management	Studying training and methodical literature
	Business-plan development
High-tech market	Analyzing management tasks on laser market
Foreign language	Work according to individual plan
	Second regular training phase (1 week)
Management	Controlling at an enterprise (firm)
	Management of a firm specializing in R&D and high-tech production
	Staff management basics
Marketing	Development of advertising campaigns for high-tech products
	Arrangement of exhibitions

Program Item	Brief contents of topics
R&D management	Peculiarities of patenting abroad
	Peculiarities of investments in high-tech
	International R&D cooperation in Europe
Legal regulation in economy	Entrepreneurship in Germany, support of small and medium businesses
High-tech market	Problems of conversion enterprises
Foreign language	Testing and communication practice (English and German)
Final lessons	Presentation of business-plans developed by trainees
	Individual interviewing

PROJECT MANAGEMENT: KNOWLEDGE AND TRAINING

Alexander YANCHEVSKIY
LETI–Lovanium International
School of Management,
St. Petersburg, Russia

Many of the most difficult engineering challenges of recent decades have been to design, develop, and implement new systems of a type and complexity never before attempted. The creation of these systems within acceptable schedules and budgets has required the development of new methods of planning, organizing and controlling events. This is the essence of project management. Management of a project differs in several ways from management of a typical enterprise. Succinctly, a project is an organized one-time effort to achieve a focused objective.

Major processes in project management include:

- identification of a need for a product or service;
- definition of the goals of the project and their relative importance;
- representing the project as a set of activities;
- definition of time and resources necessary to implement each activity;
- definition of technological order of activities;
- scheduling the project in time;
- developing a budget and financing scheme;
- integration into a complete project plan;
- monitoring the project during implementation and taking corrective actions.

Successful project management implies in depth knowledge of marketing for successful identification of needs for a new product and projections of future cash flows, knowledge of production planning and costing, inventory management and logistics, capital budgeting and corporate financing, scheduling methods (PERT, CPM), expertise in organizational design and people management. Inevitably it requires strong leadership and ability to take decisions under uncertainty and pressure.

Relevant training programmes regularly should include basics of above-mentioned functional areas integrated into a systematic frame of project management. Simulation of project planning and implementation based on one of

C.M.R. Verkoeyen et al. (eds.), Management Training in High-Tech and R&D, 151–152.

the real world software systems (Super Project Expert, Microsoft Project, etc.) is essentially important at the final stage of training programme.

The necessity for training can be explained by the very well known practical experience: a carelessly planned project will take three times longer to complete than expected; a carefully planned project will take only twice as long. Similar or even more convincing evidence is related to financial results of the projects.

PROBLEMS OF CONVERSION AT THE OBNINSK ENTERPRISES AND PREPARATION OF THE MANAGERS IN THE SPHERE OF HIGH-TECH

Igor GONNOV, Alexander SOROKIN
Association "Obninsk-Technopolis", Obninsk, Russia

The town of Obninsk was established as a town of science under target state programs in complex solving of problems related to nuclear power engineering. The state scientific sector comprises 13 research institutions, including three State scientific centers of the Russian Federation. The town has state-of-the-art production base at the instrument-making factory "Signal" and pilot engineering divisions of research institutes, as well as a significant training and re-training base.

Though originally established as a scientific and technological center of atomic industry, now Obninsk is a large multi-profile scientific center.

The main problem in the R&D sphere of the town consists in drastic and constant during the last five years reduction in financing for fundamental research with simultaneous reduction in financing for applied R&D and experimental-design work (by 20 times in comparable prices), which affects solving such a problem, as preservation of the scientific and technological potential available and jobs in the scientific complex. The main reason for the present situation occurrence proves to be the significant reduction in funding for atomic power engineering and defence topics along with the absence of really working state program of conversion and employment in the said fields.

The public Association "Obninsk-Technopolis" established in 1992 and uniting today the leading research institutions of the town, its higher educational institutions, the research park, the joint stock investment bank and a number of small innovative enterprises plays an active part in arranging and coordinating activities of various authorities and organizations, of the town community trying to contribute to the preservation and development of Obninsk as developing-in-complex scientific center, in assisting conversion processes and integration of the town's R&D potential into the system of market relations.

One of the activities conducted by the Association "Obninsk-Technopolis" along this direction was a Russian-American seminar held in Obninsk by the initiative of the Association in 1994 under a joint project of the Union of Russian Science-Towns Development, a member of which Obninsk is, and the National Defence University of the USA, which initiated the project. The project consisted of three main elements: exchange of information on the USA conversion experience, analysis of the situation in Russia at the regional level and that of enterprises and practical seminars for top management of Russian defence enterprises.

C.M.R. Verkoeyen et al. (eds.), Management Training in High-Tech and R&D, 153–158.

In 1992–94 the Representatives of the Industrial College of Armed Forces (National Defence University of the USA) Dr. Klair Blong (now a specialist of the Federal Agency of Emergency Situations Management) and Mr. James Lekky carried out research within the framework of the above project in Nizhni Novgorod, Obolonsk of the Moscow region, Obninsk and Kaluga.

In Obninsk the seminar "Problems and Ways of Conversion at Enterprises of Science-Towns" was held on the basis of the research and methodical materials published by the Department of Technological Assessment of the USA Congress. The representatives of the Physical-and-Power-Engineering Institute, Obninsk research and production enterprise "Technology", the affiliate of the Karpov Research Physical-Chemical Institute, Medical Radiology Scientific Center, Central Institute of Qualification Improvement, instrument-making factory "Signal", Obninsk joint stock investment bank "INVESKO-BANK", Obninsk insurance company "Nauka" ("Science"), a number of innovative enterprises participated in the seminar. The purpose of the seminar consisted in familiarization with the conversion experience, the techniques and strategy of conversion at American research institutions and industrial enterprises working in the fields of science and technology similar to those in which Obninsk institutions and enterprises specialize. At the seminar attention was paid to the role of banks, public R&D institutions and local authorities in carrying out conversion, training and re-training of managers and R&D specialists for work under market conditions.

Before the seminar the specialists of the National Defence University of the USA carried out the opinion poll of Obninsk enterprises' representatives (via questionnaires) on conversion problems. 40 people took part in the poll, of whom 25 represented R&D institutions, 5 – industrial enterprises, 2 – town authorities, 8 – other spheres (small innovative enterprises, banks, public organizations). The age of 75 % of respondents was 40–50 years.

The analysis of the questionnaires containing more than 30 questions allowed the American specialists to identify the attitude of the respondents to the conversion implementation at Obninsk enterprises in such aspects as understanding of general conversion problems, also proceeding from the economic situation in Russia, in the town, at the enterprise; the concept of conversion implementation; enterprises' strategy in carrying out conversion; obstacles in the conversion process; criteria of conversion success. Below we give the replies given by 50 or more percent of respondents:

CONVERSION IS UNDERSTOOD AS:

– Increase in the output of civil products – 80 %;

– Arms reduction – 60 %;

– Changes in enterprise's specialization – 50 %;

– Introduction of high technologies – 50 %.

ELEMENTS OF CONVERSION STRATEGY:

- Enterprise restructuring – 80 %;
- Search for market – 70 %;
- Maximum utilization of facilities
 and human resources of the enterprise – 60 %;
- Search for financing – 50 %;
- Specialists re-training and high-tech management training – 50 %;
- Military export, foreign market – 50 %.

CRITERIA OF SUCCESSFUL CONVERSION:

- Support of the state – 85 %;
- Full utilization of the enterprise potential – 80 %;
- Steady financing – 70 %;
- Close connection between research and production,
 fast growth of new products fabrication, new markets – 60 %;
- Specialists re-training and high-tech management training – 50 %.

MAIN OBSTACLES IN CARRYING OUT CONVERSION:

- Financial problems and inflation – 80 %;
- Weak governmental control – 75 %;
- Failure to develop conversion strategy at enterprises because
 corresponding Ministries do not have any such strategy – 70 %;
- Lack of marketing – 60 %;
- Psychology of employees, possibility of worsening
- the psychological environment of the enterprise – 50 %;
- Supply problems, as contacts with the enterprises of Russia
 and the former USSR have been lost – 50 %.

Brief conclusions characterizing conversion process at Obninsk enterprises.

1. Specific Russian difficulties in carrying out conversion, first and foremost, economic decay, slow transition to market economy, absence of the governmental and often local control.

2. Abrupt reduction in the volume of governmental orders and funding, absence of investments and market, burden of infrastructure on enterprises' shoulders.

3. Need in specialists re-training and high-tech management training.

General recommendations given by American specialists regarding R&D conversion strategy at the level of an institute or a laboratory.

1. Identification of civil applications:
   - create and train a group of specialists to find possible applications of technologies developed at the institute;
   - study civil industries where the institute's R&D may be used;
   - define as many promising fields of cooperation with industry, as possible.

2. Estimation of potential market:
   - estimate the Russian market by identifying the needs corresponding to the institute opportunities and the volume of potential demand;
   - estimate the world market via studying published materials, inviting consultants, using services of a trade agent abroad.

3. Choice of partners:
   - set contacts with the representatives of industries where the institute's technologies may be applied;
   - conduct seminars of searching character, exchange ideas with managers and specialists of industries chosen;
   - there, where interests coincide, start planning the introduction of technologies.

4. Determination of funding sources:
   - learn about the programs of the Federal Russian Government in R&D support;
   - look for partners in your country;
   - look for partners abroad via participation in trade fairs, direct contacts and correspondence;
   - prepare information on costs and prices which may be of interest to foreign partners and investors.

5. Work with customers:
   - from the very beginning work with customers, so that you could know their financial opportunities, what they think to be significant in the technology offered by the institute and what they can do without;
   - attach serious attention to any ideas of products improvement offered by the customers.

Many of these and other recommendations given by the specialists of the National Defence University of the USA may seem too evident. However, one main idea can be seen in all of them: a manager must fully control the situation for which purpose, first, he has to understand what lies on the surface of the problem approved for implementation, in particular, the problem of carrying out conversion within the framework of a certain institute or laboratory, and be professionally trained to solve further tasks connected with establishing the production of new-kind competitive civil goods and entering the Russian and international markets.

The seminar resulted in signing protocol of intent between the Industrial College of the National Defence University of the USA and the Association "Obninsk-Technopolis" on further cooperation in carrying out conversion via arranging similar seminars, exchange of information, presentation of these organizations' activities in both countries. By the opinion of the American specialists the cooperation of this kind will allow the governments of both countries to more efficiently solve conversion problems which have much in common and will allow specialists to better understand the strategy of their enterprises development which can make the complicated process of economic transformations easier both in Russia and in the USA.

When conducting the seminar the American specialists made an assumption that establishing of the Association "Obninsk-Technopolis" and an innovative company rendering professional assistance in innovations and new technologies transfer will contribute to the efficient implementation of the R&D potential of institutes and enterprises entering the Association, will make them more independent in their technologies promotion under transition to market economy. In October 1995 Obninsk administration, the Association "Obninsk–Technopolis" and the State Foundation for Assistance in Small High-Tech Enterprises Development set up Business-Innovative Center for the development and coordination of innovative activities in the town and the region as well as for the representation of R&D institutions' interests in the implementation of their R&D and industrial potential.

The main directions in the Business-Innovative Center operation are high-tech innovative activities connected with the support of short-term innovative high-tech projects for the implementation of which today R&D institutions lack centralized funding, but which can be implemented by small enterprises. The Center also selects and transfers defence technologies to civil industries with establishing high-tech business and commercial cooperation on this basis. For this purpose the Center arranges training, re-training and exchange of specialists in high-tech management training.

To develop the last direction Business-Innovative Center of the Association "Obninsk-Technopolis" prepares the representatives of Obninsk enterprises for the participation in the special American program of business training (SABIT) organized for the specialists of defence industry and officials of local administrations from Russia, Ukraine, Bjelorussia and Kazakhstan. The program implies 2-months practical training in the large centers of the USA defence industry which are successful in carrying out conversion and have intentions to collaborate with similar enterprises on the territory of the former USSR. The practical training includes 2-weeks course of business management in Washington and six weeks of practice in conversion and production reorientation at private enterprises. The purpose of the program consists in introducing the participants to the peculiarities of governmental company management under transition to market economy.

The program for enterprises managers is focused on training in business and finance planning, new products fabrication and sales, organization of system for providing the products users with services. The program for the representatives of local administrations is concentrated on the problem of establishing favorable conditions for carrying out conversion at defence industry enterprises.

The task of management training is solved in the town by the partners of Business-Innovative Center of the Association "Obninsk-Technopolis" within the framework of the joint business cooperation aimed at the development of entrepreneur activities in different, first of all, innovative fields in order to train, improve qualification and re-train managers, to support entrepreneurship and introduce market relations to various spheres of life. Here is the list of these organizations: Central Institute of Qualification Improvement for Top-Managers and Specialists of the Ministry of Atomic Power Engineering of the Russian Federation; International Academy of Modern Knowledge of the State Committee for Higher Education of the Russian Federation; non-governmental higher educational institution – French-Russian Institute of Business Administration and established on its basis Training-Business Center of "Morozov project". Thus, the training programs of the French-Russian Institute of Business Administration have been developed proceeding from the experience of management and marketing training by higher educational institutions of the Federation of European Schools of Russia, France, USA. The educational program of the Training-Business Center of "Morozov project" is based on more than 30 courses and seminars in management and marketing developed by the Academy of Management and Market (Moscow). International Academy of Modern Knowledge conducts its educational activities in professional training along the directions "Management" and "Marketing" according to the training programs of the Open University of Great Britain with the use of the remote training system.

At present the joint work of the Business-Innovative Center of the Association "Obninsk-Technopolis" and the above training institutions is aimed at the adaptation of the training programs and their introduction for re-training of specialists from Obninsk conversion enterprises and high-tech management training.

PROJECT AND QUALITY MANAGEMENT TRAINING WITHIN THE FRAMEWORK OF THE PROGRAM "SUPPORT OF SMALL BUSINESSES AND NEW ECONOMIC STRUCTURES IN SCIENCE AND SCIENTIFIC SERVICES FOR HIGH SCHOOL"

Alexander VICTOROV
St.Petersburg Electrotechnical University, Russia

The program "Support of Small Businesses and New Economic Structures in Science and Scientific Services for High School" should ensure innovation activities development in higher educational institutions of Russia and these activities integration into similar activities conducted in the industry of this country and abroad. The program is built by a regional and subject principle. In the regional aspect it has twelve regional sections and in the subject one - six subject directions:

- innovation projects;
- new economic structures;
- specialists training;
- information dissemination;
- international cooperation;
- legal support of the program.

The program management and its scientific and methodical support implement the program proceeding from the integrity of the said regional and subject principles.

The program tasks are fulfilled under specific conditions of the Russian economy transformations. The main peculiarities of these conditions are as follows: lasting economic recession and, correspondingly, recession in innovation activities; non-payments; hidden inflation of ruble and dollar, as well as rather intensive growth of prices for power supplies, of rent and other expenses of enterprises conducting innovation activities. The said factors, as well as some others, strongly influence the small businesses choice of their strategy in business-planning and high-tech projects management. In this connection great importance is attached to the retraining of project managers and quality managers proceeding from the above factors and the foreign experience accumulated under the conditions of steadily developing economy. Besides getting professional knowledge in high-tech business, project managers should become aware of the fact that this sphere of business is not and within the next fifteen years will not be highly profitable and that under the present economic conditions in Russia the possibility of bankruptcy in this sphere is higher that anywhere else. As a rule, mentally project managers are not ready to perceive this specific feature of Russia and make economic mistakes when they formulate the strategy of their business

C.M.R. Verkoeyen et al. (eds.), Management Training in High-Tech and R&D, 159–163.

development and aim at obtaining maximum profit. Besides, the major part of small businesses in science and scientific services do not look for their market segment, but just try to duplicate industrial enterprises and research institutions on the market of high-tech products and scientific services. This is a serious mistake which often entails bankruptcy of small businesses. One of the tasks in specialists re-training for this business sphere consists in training managers of firms and projects to start their business with market analysis and the choice of their market segment.

At the first sight arranging high-tech business under innovation activities recession may seem an unnatural task having no prospects. It is the very thing that happens in practice, if small high-tech businesses are not provided with serious governmental support. There arises the question: "Why should the state provide support to high-tech businesses?" Under transition to market economy with certain governmental regulation small businesses in science and scientific services fulfill the following very important functions:

- ensure additional jobs for highly qualified specialists preserving the scientific and technological potential;
- play the role of interface between research institutions and industrial enterprises;
- act as "discoverers" on the market of high-tech products and scientific services studying this market by the example of small pilot batches of high-tech products. The results of such market analysis are important for the future commercial production of these products by industrial enterprises.

Within the process of specialists re-training it is very important to train specialists in arranging the support for businesses dealing in science and scientific services at the federal and regional levels, on one hand, and to train managers of enterprises and projects in the rules of getting this support, on the other hand.

The problem of non-payments faced today by the country has become the main one for small high-tech businesses. While big enterprises have some reserve of current capital for temporary compensation of non-payments, small businesses, as a rule, have no such reserve, therefore, non-payments make the main reason for the bankruptcy of the significant part of high-tech small businesses. Non-payments of high-tech products customers also strongly affect the ability of enterprises to use the developed infrastructure. In this connection the problem of timely arranging training and consulting for small business managers and project managers in the ways to compensate non-payments at the regional and federal levels becomes most vital.

Products certification and activities licensing make a specific problem in Russia. In this country the services in products certification and activities licensing are rather expensive. That results in the fact that small enterprises having no money for certification and licensing are looking for the ways to avoid requirements in certification and licensing. Therefore, it is very important to correctly put accents in training specialists involved in establishing infrastructure for business support and to persuade them in the necessity of implementing regional and federal policy of certification and licensing for small enterprises in science and scientific services, as well as for their products.

In arranging re-training of high-tech business specialists the main attention is attached to re-training project managers of three levels:

- managers of regional and subject sections of the program;
- managers of small businesses and new economic structures in science and science services;
- managers of concrete projects.

Contrary to the majority of Russian educational institutions, including those with the participation of foreign specialists, the said program is intended to train managers for concrete projects implemented under other divisions of the program. Such an approach ensures the increased interest of trainees to the training offered and the increased efficiency of the training itself.

Organizational and methodical structure of specialists re-training within the framework of the program is based on two principles: ensuring the educational and methodical integrity of training activities in all regions and arranging re-training proceeding from specific features undoubtedly available in the regions. The main training and methodical tasks of the program are basically solved by three organizations which attract specialists from the regions. Fig. 1 provides the scheme of these organizations interaction in preparing educational and methodical material.

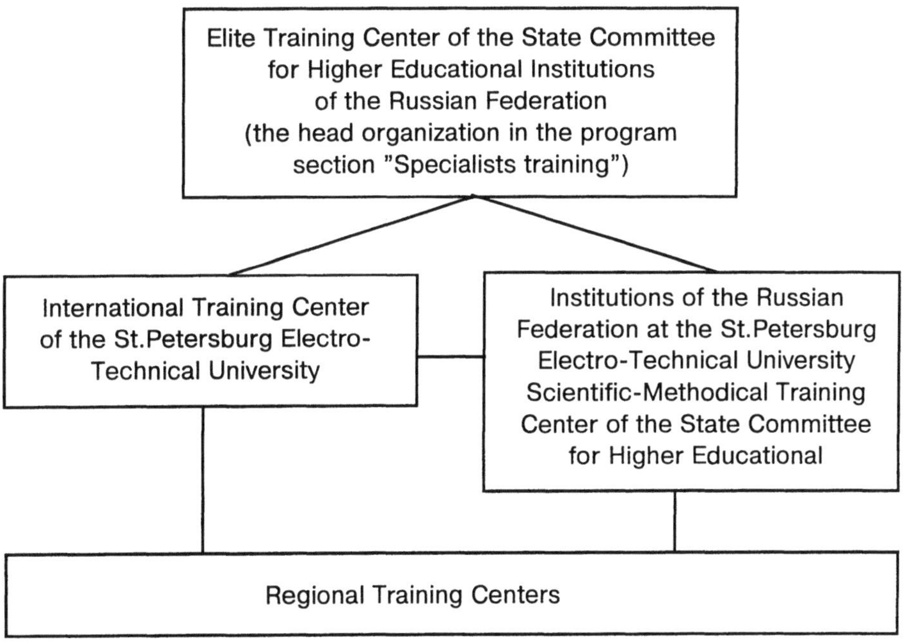

Fig. 1. Organizational Structure of Preparing Educational and Methodical Materials for the System of Specialists Re-Training.

In 1995 the organizational structure of arranging and conducting seminars for specialists re-training according to the program was established within the framework of the program. It includes the Elite Training Center – the head organization in the corresponding section of the program and Training Centers (TC) in the regions. The said structure is shown in Fig. 2. It should be mentioned that Elite Training Center and the major part of regional TC's were established as organizations of the State Committee for Higher Educational Institutions of the Russian Federation which ensures the solution of problems connected with their training activities licensing. At present a single data base on educational services which are or may be rendered by regional TC's is being formed. Due to this data base it will become possible in the nearest future to draw up a necessary plan of training to match a certain problem and to ensure the implementation of the said plan by training staff from all economic regions of Russia.

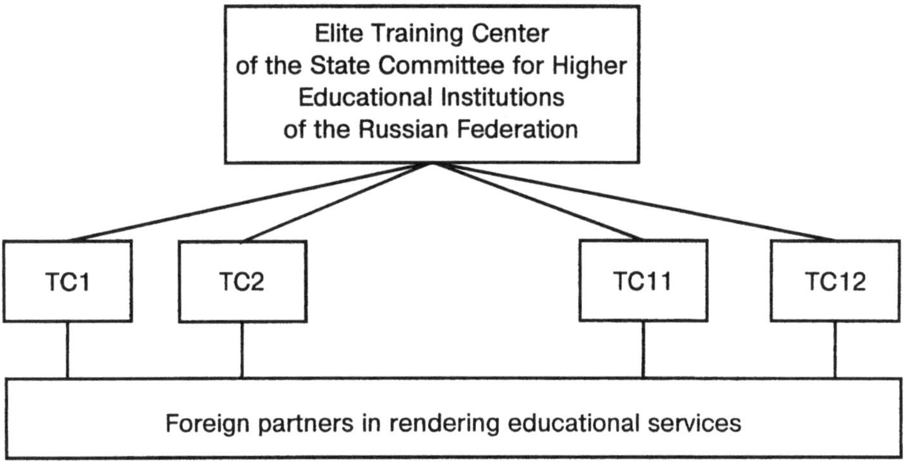

Fig. 2. Organizational Structure of Specialists Re-Training

A computer text-book being gradually developed according to the plan approved makes the basis for the formation of the sets of educational and methodical materials. The advantage of the computer text-book in comparison to conventional printed text-books consists, first of all, in the possibility to promptly make changes and addenda caused by changes in legal documents at regional and federal levels. Besides, the computer text-book provides an opportunity to vary the depth of material studies beginning from general considerations to concrete legal documents given in the Appendices. It should be also mentioned that Appendices to the text-book are active and users may work with them like with regular files, preparing working documents included. The system of information exchange between the regions implemented under the program section "Information dissemination", electronic mail included, provides for timely updating of Appendices to the text-book. In December 1996 a seminar for the managers of the program regional sections and managers of regional small businesses will be conducted by a frontal method within the framework of the current training system. Groups of trainees will be formed by the regions and training will be arranged in the regions on a single topic with the use of common educational and methodical material. Besides, a teleconference is planned. Mastering such a method is of principle importance for overcoming financial difficulties occurring today when a specialists has to go to another city to take part in a seminar.

In conclusion it should be mentioned that people in charge of developing the system of specialists re-training in project management and quality management under the program "Support of small businesses and new economic structures in science and scientific services for high school" are striving to establish the integral spatial educational system able to timely react to changes in the economic situation and legal base at federal and regional levels.

MANAGEMENT DEVELOPMENT: SUBJECTS AND LEVELS

Liliane VAN HOOF
University of Antwerpen (UFSIA), Netherlands

A university degree and/or several years of experience are strong trump cards in the business world. However to build a succesfull management career adaptations to new demands and mastering of new skills are necessary, as are delivery of top prestations in the own specialist field as well as in the general management area. Not only needs someone to manage several projects at the same time, most likely in a team of specialists, he needs also to be aware of the environment in which these activities take place. This environment is changing constantly: "nothing is constant but change". Flexibility and adaptation of management is constantly requested.

This contribution places management learning and development in the business environment and strategic process with special linkages to the realities in Eastern Europe and the Russian Federation. It analyses the training philosophy, contents and levels of management education and the training process. Ultimately some recommendations are brought forward. Within the range of management development this contribution gives a first impression and a broad overview.

1. Sources of change

The strategic process (see Fig. 1) puts the emphasis on analysis of the environment, the companies capabilities and aspirations in order to formulate the strategic orientation of the enterprise. Change can have many sources and can be initiated by the slightest adaption of any element in this strategic process. An exhaustive analysis of all these different sources would be too vast of an undertaking however some important issues in the environment need to be highlighted.

1.1. Globalisation is leading to a global village.

This globalisation is due to deregulation; new trade zones (NAFTA, APEC,...); new emerging markets and the strong economic performance of South-East Asia; the opening of Central and Eastern Europe; World brands (Benetton, Coca Cola, Sony,...); transnational companies; diminishing of trade barriers (WTO,...). Globalization will change dramatically the relations between countries and continents and influence the companies activities.

C.M.R. Verkoeyen et al. (eds.), Management Training in High-Tech and R&D, 165–172.

1.2. Rapidly changing technologies with special reference to the information technology.

"Technological change is often difficult to predict, as are the time and costs required to develop and commercialize new technology. In many areas of technology, particularly leading-edge technologies, there are significant disagreements between the experts about which of several competing technologies will work. Small firms lack the resources to develop competing technologies and usually have to bet on one" (Slatter, p. 9). This puts the small firm at a tremendous strategic risk.

A special place is taken by information technology that forms the basis for new industries (internet, information superhighway,...); increases the economic value of information (scanning methods at the cash-register in order to control inventories, gives an analysis of the buying behavior of customers,...); virtual networks are possible in a lot of activities as pay for production, electronic sales,... ; international standardization of messages.

This technological revolution has to be followed closely in order to stay in the market place, e.g. large companies take the bulk of deliveries from companies hooked up to their system via EDI, outdated products loose market share.

1.3. The consumer demands are increasing, mass-customisation is requested and possible thanks to the information revolution, customer preferences are changing quickly and product life cycles are getting shorter. (e.g. average life cycle of PC: 6 months, cars 4 years,...). The consumer is the one who ultimately orders the goods and pays, so he will request adaptation of the producers: consumer sovereignty is getting more absolute.

1.4. Throughout history (Stern, p. 30-54) there have been various managerial and organizational practices (schools) in vogue to fit the demand of the times. A total explanation of these diverse schools would go outside the scope of this contribution but a short overview of twenteeth century practices is revealing as to change: the classical school with Weber's bureaucracy and Taylor's scientific management; the behaviourial school with emphasis on human relations viz Maslow's knowledge of needs and theory X and theory Y from Mc. Gregor and organizational behaviour; the quantitative school with operations research and industrial statistics; the management function school: analyses the job of managing as planning, organizing, commanding, coordinating and controlling. In this respect Fayol's 14 principles of Management should also be stressed together with the managerial roles of Henry Mintzberg: interpersonal roles, informational roles and decision roles. One of the the latest big schools is the total quality management approach.

In reaction to envrionmental changes the above mentioned management schools came into existence. The field of management and consequently management education is continuously changing due to feedback loops and concertations with reality. The principles that were more frequently used during the last ten years and revolutionized management practices in several companies apart from the above quoted total quality management are: business process reengeneering (BPR) best practices (tools: benchmarking, awards, review of BPR,...), just-in-time management (JIT), flat organization, empowerment of employees, team management, downsizing, linear modelling (LM)...

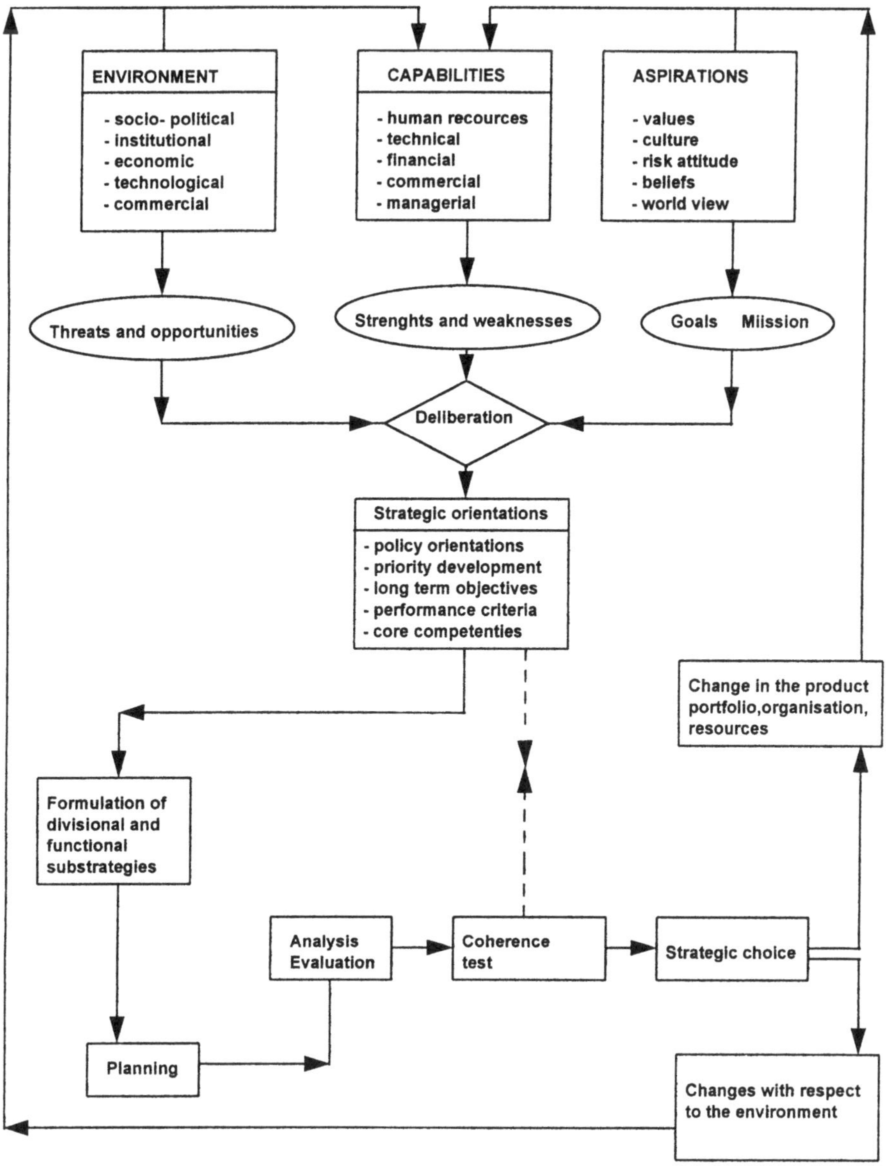

Fig. 1: Strategy Formulations

Conclusion: management theories and education are a reflection of the era and a response to environmental changes and insights.

1.5. The generic strategic responses of companies to the challenges posed by the environment and reflextion added through capabilities, values, management theories and education have been:

1) Mergers and acquisition to increase the size of the company and the market power.

2) Concentration on core activities and divestment of non core activities to suppliers (focus strategies)

3) Optimization-strategies (BPR, TQM, JIT, LM,...) (flexibility strategies)

From these three generic strategies the first two are in the long run quite irreversible and possibly most promising.

1.6. The above mentioned generic strategic responses of companies are influenced by the industrial life cycle and the key success factors that evolve over time.

In the introduction phase technology and product features are of the utmost importance to create advantages and beat the competition, in the growth cycle flexible manufacturing is required whereas reliance on core-competencies and customer focus and service are used frequently in the saturation stage. In the decline stage the discernible key success factors are: embedded management style, vision and people orientation. Fig. 2 places several industries on the life cycle together with the companies reaction along the life cycle diagram.

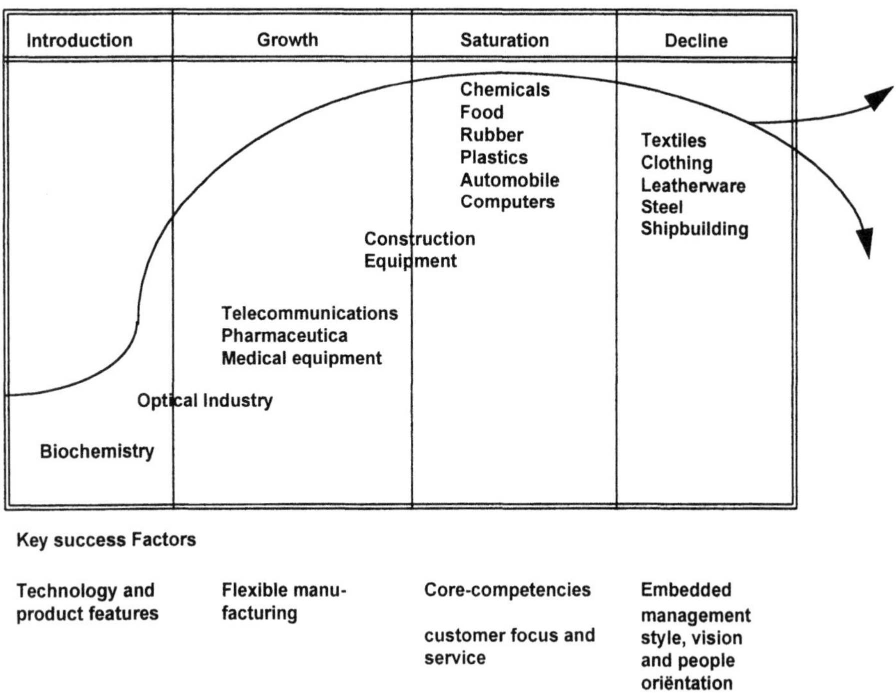

Source: Arjan J. Van Weele – Ondernemingen en inkopen in Wild Water, IPO Management School, 15 November 1996.

Fig. 2. Critical success factors depend on the evolution of the industry

Conclusion: A longitudonal studie of industries indiciates changing key successfactors. This should be reflected in the management of the company.

2. Age of discontinuity

In Central and Eastern Europe change started slowly after the election of M. Gorbachev in 1985 but progressed rapidly after the fall of the Berlin Wall in November 1989 and the liberalization of prices in January 1992 in Russia. This change was a revolution on the political, ideological, economic, and cultural field. Schematically the reforms can be identified as in Fig. 3.

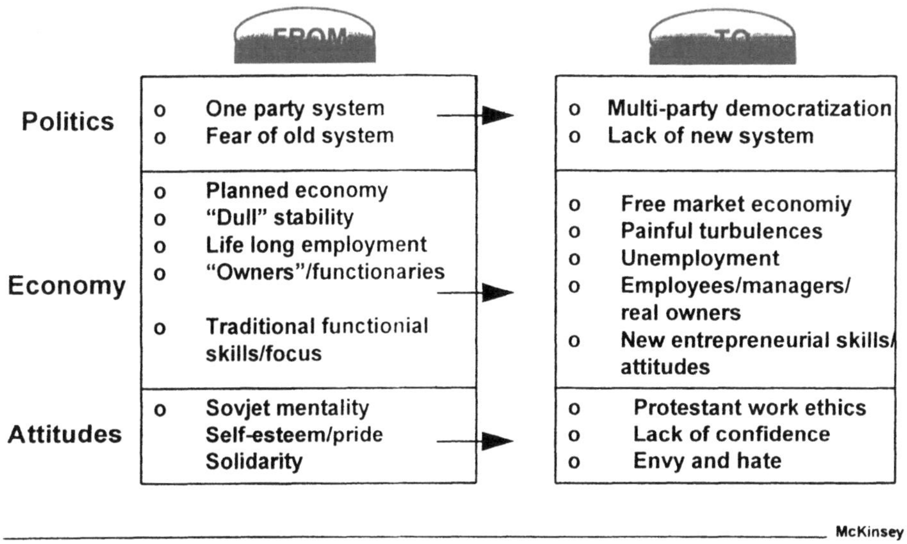

Source: M. Peter Krajlic, Mc. Kinsey & Company, Prague October 1996, Ceeman Conference.

Fig. 3. Structural change in Eastern Europe

A new set of rules and parameters are introduced, the macro-economic environment and the regulatory system for business are adapted beyond recognition. Such dramatic changes demand a lot of flexibility from the institutions and the individuals since new techniques, values and tools have to be learned and used overnight and several gaps with the rest of the world have to be closed (technology, productivity, motivational, management & marketing, capital, legislation, democracy and environmental) or as Peter Krajlic from Mc. Kinsey states: the gaps reflecting the heritage of the communist system have to be resolved (Fig. 4).

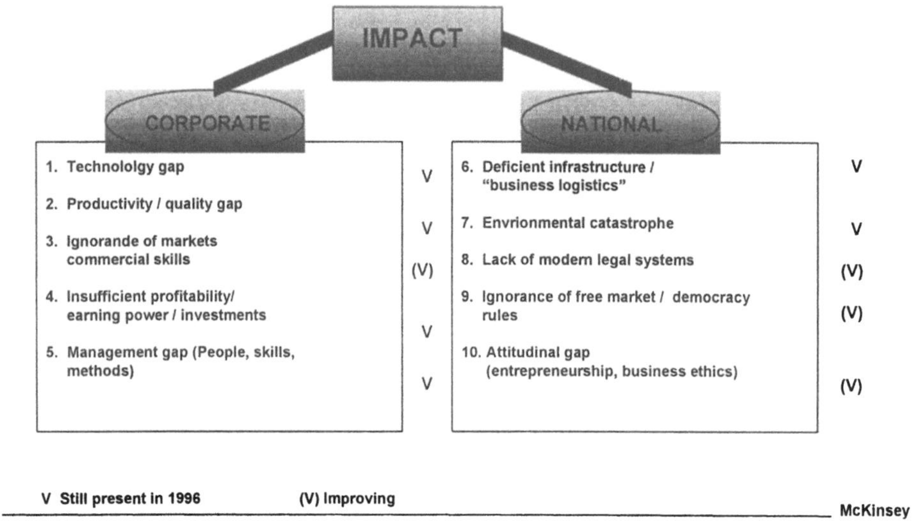

Fig. 4. Gaps as a result of the communist system

This radical evolution creates a new paradigm for management, because in order to survive as a company or an individual innovation and creativity is needed to respond to these challenges. Learning is the best advice as well as stimulation of the entrepreneurial spirit and flexibility.

Changes in political and economic systems can be introduced relatively fast, however changes in value systems will take a longer time depending on the population specifics, location, religion and beliefs.

As an illustration we include figure 5 with the differences in value systems between Management in East and West as compiled by J. Prokopenko (1992).

These differences are generally observed and do fluctuate from individual to individual.

DIFFERENCES IN VALUE SYSTEMS BETWEEN MANAGEMENT FROM EAST AND WEST

	WEST	EAST
Importance Attached	Wealth Informality and Competence	Equity Protocol, Rank and Status
Emphasis on Motivation	Individualism	Group Unity
Education as investment	Personal Development Success	Prestige
Dealing with Conflicts	Creative Energy to be managed	To be avoided
Approach to Output	Quality, Sales, Marketing	Production, Quantity
Communication	Horizontal/Vertical, Open, Networking	Vertical, Closed
Responsibility/Risks	Looking to take Risks	To be avoided

Source: J. Prokopenko, "Human Resources Management in Economics in Transition: The East European Case", MAN DEV/66, ILO, 1992

Fig. 5. The differences in value systems between Management in East and West as compiled

In the next decade these values will move closer together and certain individuals and certainly the younger generation will adhere to other norms in business compared to their predecessors.

These discontinuities unleash an unparallelled amount of energy and creativity and will increase the prosperity of society, the indivudual and companies over time. In the meantime however sacrifices have to be brought but the picture is not all that bleak because of the presence of natural resources and an able and educated population: a high percentage of scientific workers and engineers are strong competitive advantages in a global world. If this can be combined with work ethics and management learning and a judicial industrial policy (away from the military and heavy industry) the future looks promising indeed.

3. Training Philosophy

Organisations, individuals and societies (Ph. de Woot, p. 32) must adopt a continuous learning philosophy so as to create a learning society, learning organizations and learning individuals. The old adagium: "There are more opportunities through learning" still holds.

This learning society means learning from cradle to grave and this requires completely new educational infrastructures. In management education this means creating less managers and more entrepreneurs. It also means to create citizens with a set of values, the capacity to learn, to adapt, to face turbulence, to live with complexity, to have a greater capacity to communicate, to work in teams and to be more internationally oriented. The concern is not to create one-dimensional managers. Thirdly a strong education chain has to be built in the learning society, no weak links are tolerated and as to management capabilities we must stress the leadership skills, values and entrepreneural capacities in the education process.

A learning individual needs to "know himself better", he needs continually to challenge and adjust mental models through action and critical reflextion processes and any deficits encountered in content knowledge, self knowledge, diagnostic skills, application skills, teamwork skills and self control should be rectified. This implies a continuous feedback loop between action and reflextion, information dissemination and gathering, studying and perfection of one's skills and utilizing these skills in the enterprise.

A learning organization needs to adapt a system's thinking approach to understand complexity, to focus on teams and needs to give workers the possibility to work together at a common cause/project in an open constructive climate. An audit of learning needs should single out deficit areas in the organization and they have to be rectified (infra).

These reflections on learning and the business environment that obviously changes (supra), requires adjustments from organizations and new skills from individuals as (Angehrn, p. 24):

- being able to operate internationally across continents and have a good understanding of cultural differences;
- select essential data, interpret signals, determine clear objectives and act rapidly, often on the basis of excessive though incomplete information;
- be flexible and adapt continuous learning as a philosophy of life;
- get things done WITH others.

How is this philosophy reflected in the concrete skills managers need to obtain?

4. Managerial competencies

A variety of skills are associated with effective managerial performance. A combination of technical, interpersonal, conceptual and diagnostic skills is necessary to manage efficiently and the skills needed vary according to the position in the company and over the lifetime of a manager as R.L. Katz (1974) and Sthal (1995) indicate.

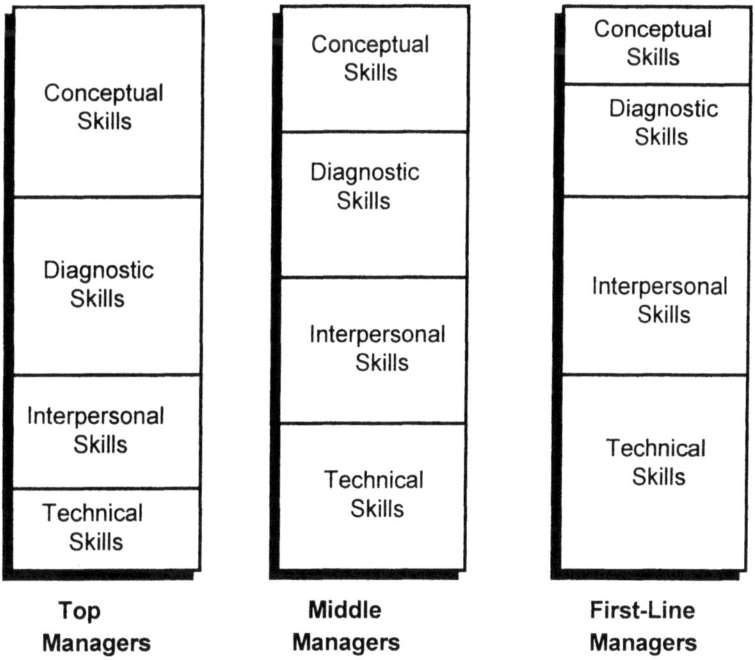

Fig. 6. The importance of Management Skills by the Organizational Level

4.1. Technical skills.

They consist of specific competences to perform particular operational tasks and include proficiency with the methods, procedures, techniques, knowledge and equipment involved with the specific functions of viz manufacturing, sales, accounting, finance or engineering.

Technical skills required in the management field include:

General management. The tools of strategic analysis and planning (see Fig. 1 for the strategic process).

Accounting, auditing and finance. In order to be able to run a business financial fundamentals as cash flow, cost allocation, NPV, IRR, valuation methods, ratio analysis, asset management, risk management, budgetting, direct costing, contribution analysis, investment appraisals, break even analysis... have to be understood.

To retrieve the data to apply these techniques an accounting system needs to be installed.

Human Resources Management. This includes the following main fields: man power planning and development, recruitment and selection, career management and people development, decruitment; integration, motivation and involvment. Jobs engineering, leadership cycle, compensation and reward systems, information, communications, industrial relations, social laws and legislation, taxes on wages,... The gap between the traditional and new practices and requirements in pay and compensation, and performance management are quite extensive as enumerated in figure 7 and 8 (Poor 1995). The job of the Human Resources Manager will have a completely different content from that under central planning and will be a fascinating job.

Traditional Pay Practice	*New Pay Practice*
- Strong Government Control - Egalitarian - Grading Structure Based on Hierarchy - Pay for Status - Pay for Quantity - Low Base and High Incentive Ration - Tax Free Benefits - Separate Pay and Personnel Management	- Reduced Government involvment - Non Egalitarian - Grading Based on job evaluation - Pay for Performance - Pay for Quality - High Base and Moderate Incentive Ratio - Government Taxation of Benefits - Pay Integrated into Personnel Management Practice

J. Poor - Human Resources Management in Central and Eastern Europe, Ceeman 1995.

Fig. 7. Pay and Compensation

Traditional Practice	*New Requirements*
- Tayloristic bonus and variable pay	- Variable pay for economic and competency base targets
- Training was merit	- Training becomes an integrated part of Personnel Developments
- Specific appraisal for managers (e.g. political view)	- Same approach for the whole population of employee

J. Poor - Human Resources Management in Central and Eastern Europe, Ceeman 1995.

Fig. 8. Performance Management

Marketing. How to put products on the market to satisfy the needs of customers. The customer is sovereign.

Includes: market research, product innovation management, price and communication policy, analysis of distribution channels and logistics. Feedback between market and production is of the utmost importance. Manufacturing of products which are wanted by the customer after being production-driven for several decades demands an adaptation by all parties involved.

Law for business. A strong legal system guarantees the equality and rights of all partners in the economic process and puts the boundaries between acceptable and unlawful conduct. Enforcement of these laws is of the utmost importance as is the impartial judgement of the courts. Division of power between legal, executive and judicial branches is a conditio sino qua non for a market economy to develop.

On a decentralized level all players have to accept the rule of law as a guidance for their transactions.

Several laws should be mastered by the manager viz: contract law in order to sign contracts nationally and internationally, company law, fiscal and social law, laws in connection with licensinig, patent law, trade mark protection, intellectual property, anti-monopoly law...

Production operations management, engineering. Generally production and planning was a traditional strong hold in Russian companies. However innovation management and flexibility should be introduced together with judicious use of scarce resources and less stockpiling of imputs and outputs. Warehousing stocks is expensive and stocks should be kept to a minimum (cfr. JIT management)

Information technology management. Understanding the role of the IT infrastructure, the information system's life cycle, IT development and innovation, integrating the infrastructure into the organization's management. Evaluation of investment in IT infrastructure.

How to exploit this IT infrastructure for management decisions, for innovations, for developing quality information systems, for business process, re-engineering,... This does imply computer literacy and knowledge of application software.

International business and management. Cross-border transactions in the form of transfer of know-how, licensing, export, import, investments,... are increasing. In order to operate internationally, awareness of foreign cultures, negotiating tactics and foreign languages are requested. The Russian Federation posessed very efficient foreign trade organizations and highly skilled individuals, however with the dismantling of the foreign trade monopoly, the demand for international managers rose dramatically since every company can organize and penetrate foreign markets.

Business ethics and deontology. Decisions concerning profit and loss, dismissing of employees, the environment, waste management, poisonous products, ehtical products, fair business practices,... have to be made. In a society where notions as "po blatu" and "blatnoj" were the usual way of doing business an ethical background is absolutely required for business transactions and to bestow on business the respect it deserves.

Humanities background. Sociology, philosophy, geography, history,... as far as these subjects are relevant for business transactions.

Languages. In order to be able to trade, cooperate, invest internationally.

4.2. Interpersonal skills.
They constitute the manager's ability to deal effectively with and through other people inside and outside the organization and to work effectively as a team member. These skills are demonstrated in the ways the manager relates to other people - by leading, coordinating, motivating and communicating.

4.3. Conceptual skills.
They involve the manager's ability to think abstractly, to have a wholistic view of the company and to comprehend the interrelationships between the various segments of the organization. They observe opportunities where others do not.

4.4. Diagnostic skills.
The ability to understand cause and effect relationships and to recognize acceptable solutions to problems.

These skills are perceived as of the utmost importance in upper levels of the organization, but are getting through empowerment more important at the lower echelons as well.

Apart from the above mentioned competencies there are competences that can make a difference in managerial succes as: leadership (the ability to influence people to accomplish important tasks); self objectivity (the ability to realistically evaluate one's skills, values and motives as applied to a job); analytical thinking (the ability to integrate, interpret and explain patterns in available information); behavioural flexibility (the ability to modify personal behaviour to reach a goal); oral presentations (the ability to clearly express one's ideas in oral presentations and groups); written communications (the ability to clearly express one's idea's in writing); personal impact (the ability to create a good impression and instill confidence in the eyes of the others); resistance to stress (the ability to maintain work performance even under stressfull conditions); and tolerance for uncertainty (the ability to maintain work performance in unstructured and uncertain situations (Schermerhorn, 1993, p. 26).

As to general management wether businesses succeed or fail depends on the managers and their right mix of skills. The need to build understanding, commitment and skills to deal as a manager proactively with new situations demands new requirements.

These requirements can be summarized as the "Ten commandments of General Management" (Gottinger 1995, p. 31).

1. Develop a global perspective;
2. Improve strategic thinking and acting;
3. Actively focus on communicative vision and objectives;
4. Pursue continuous management of change;
5. Acquire cross-functional skills and knowledge;
6. Be able to manage more diverse resources;
7. Seek continuous improvement in quality, costs, processes and products;
8. Be able to manage creativity;

9. Do understand information and decision processes;
10 Pursue social, ethical and ecological competence to achieve harmony between innovation processes, social change and environmental quality (sustainability).

Refering to Tom Peters (management is doing), Michaël Porter (managing is thinking), we can conclude together with Henry Mintzberg (Mintzberg, 1994, p. 11) that a manager need to be "well rounded": he needs to possess the capabilities and acquire the skills to manage by information, manage through people, manage through action inside and outside the unit (company).

The manager needs to be the Integrating force in the company.

5. Sources of supply of management development

An ennumeration and short explicitation of the different supply sources illustrates the diversity of the products on the markets.

5.1. Undergraduate degree in business, economics, applied economics, commercial engineering.

This is offered in most universities and the required time of study oscillates between 6 and 10 semesters. This is a very broad education in the liberal arts, languages, functional domains of management, leadership capabilities... Stimulates the analytic and diagnostic skills of the participants. As broad an education as possible is offered.

5.2. MBA. Master of business administration: generally 2 – 4 semester full time studies. In most of the cases the participants have prior work experience and possess undergraduate degrees. Integration of business skills and previously acquired knowledge.

5.3. EMBA. Executive Master of Business Administration.

MBA designed for practising managers. The courses are offered during the weekend and in the evening in a two year time frame. Advantage is that the managers stay on the job and acquire the necessary managerial competences including the functional skills.

They can apply the acquired knowledge directly in the workplace. Participants to these courses come from all disciplines (engineering, exact sciences, philology...) and are by their companies destined for upper level managerial functions.

5.4. Management training institutes: open courses.

Several open courses are on the market that range from a basic to a very advanced level. From short, one day programs on a specific topic to 150 contact-hours programs on specialized items as e.g. real estate management, computer audit, network management, VAT topics, industrial marketing... Workshops in several fields are also available. EFMD groups several European Management institutes that offer a wide range of programs.

5.5. Open University/distance learning.

In the computer era distance learning will become more important. Networking and communication can be between computers which are close together or far apart: it may take place synchronously or asynchronously. This interaction can take place through local Area networks (LAN), Wide area networks (WAH) or through Internet. This may lead us to the virtual business school. Traditional distance learning however plays a major role because an individual can master functional skill while staying on the job even in remote areas. Underutilized branch of management education.

5.6. In-house training.

Two options are available:

Option 1: The company has its own training department and offers a wide spectrum of courses for its employees. All big companies delever a kind of an in-house training that covers the specific functional, operational and management development needs of the company e.g.

Alcatel Bell, Agfa Gevaert, TRW, GE, Siemens... Some call it even their business University. As an example the training programme of GE is explicitated in Fig. 9.

Option 2: the hiring of outside consultants.

Proprietary knowledge is taught on the market when it is lacking intracompany. Specially designed courses are offered in the widest management disciplines but are tailor-made for that specific company. Interaction and learning experience are rewarding and solve the company's needs: as to acquire certain skills, to streamline the organization, develop capable managers...

5.7. Business Innovation Centers, Incubator centers, Technoparks.

These institutions can be linked to Universities or institutes of higher education, to facilitate the interaction between science, applied research and producing marketable products. Apart from the feedback between science and practice, the possibilities for in-depth speciality discussions on innovation management, on licensing problems, technology management... are offered by inviting knowledgeable persons.

These centers are a microcosmos of mutual inspiration by the diverse participants and could trigger the discovery of same revolutionary products. Entrepreneurs get a real opportunity for a good start.

At Crotonville there is a core curriculum of courses which is focused on critical transition points of an individual's career (see Fig. 1). If you start at the buttom, there is a program that lasts two-and-a-half days called the Corporate Entry Leadership Conference. It is for all new hires coming from college campuses into General Electric; it focuses on the mission and the values, a little bit about careers, and on what you can expect from General Electric as a new employee.
A second course, called the Professional Development Course, is an elective week-long program tailored for the three-to five-year employee. It focuses on career planning and development as well as understanding the financial measurements and the thirteen businesses' products and services.

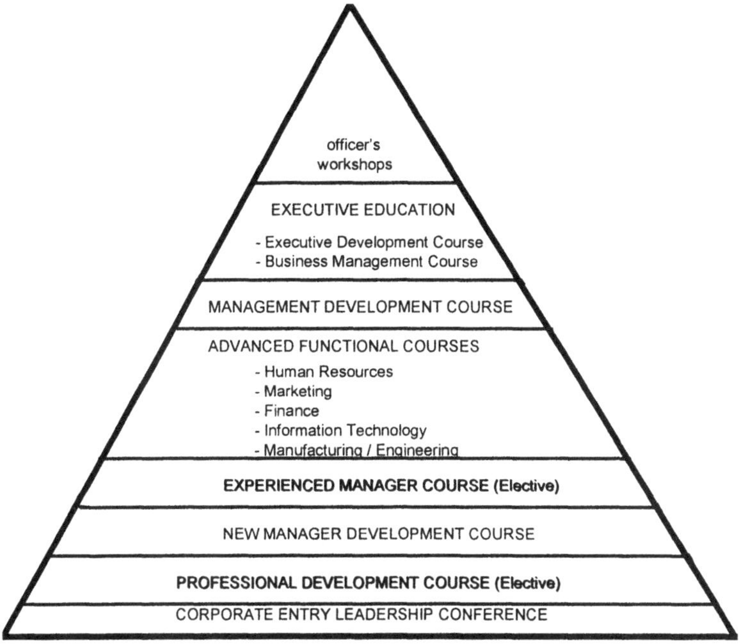

There is a second mandatory course, titled the New Management Development Course, which all individuals take within the first six months of the first time they manage professional employees. This course, in addition to being taught at Crotonville (and it is taught about 40 weeks during the year), is also taught at locations around the world. The course focuses on skill building, policies and procedures, administration and problem solving.
The next elective course in the development cycle is an Experienced Managerial Course. We like to see managers enroll in this course about five to seven years into their managerial experience. This course focuses on giving people feedback about what they do well and about the things they need to improve. There is an internal survey instrument that individuals give to their peers, subordinates, and immediate manager. Data is collected and tallied by an outside firm and is used as the basis to provide people with feedback about their managerial capabilities.
Additionally, there is a series of advanced functional courses. Each of the key functional areas of GE has an advanced level course that is two weeks to three weeks in length.
We also have three month-long courses in the executive development sequence (Exhibit 1). They each have similar themes - competing and winning in the global environment, leadership development, and developing strategic capability, but they approach these themes from different angles. The first one is called the Management Developing Course which is for high-potential individuals who have been identified through the succession planning process as having the capability to become a functional or a multi-functional general manager. The second is our Business Management Course (BMC) which is offered three times a year, and one or two of those sessions are offered outside the United States. The Business Management Course targets individuals who have the capability of advancing to a Vice President level in the corporation. This course focuses on global issues impacting GE businesses. There is a strong customer emphasis in this course as well.
The Executive Development Course, the final one in the sequence, is attended by invitation of the Chairman and builds on themes from both BMC and MDC.

Source: Keller Deborah A., Building Human Resource Capability, Human Resource Management, Spring/ Summer 1992, Vol. 31, No 122, p. 99-101.

Fig. 9. Case study: the General Electric Training Program

6. Training process

Training is a set of activities that produces learning opportunities to acquire and improve job-related skills. This applies both to the initial training of an employee, and to upgrading someone's skills to meet changing job requirements. The best climate for training is where people want to learn and are reinforced for learning.

The identifiable steps in the training process are aligned to the strategic process and assess the strenghts and weaknesses as to human capital and the way it ought to be changed and improved. It searches for training opportunities outside the firm and evaluates the threats if no training takes place.

1. Determine needs of the organization.
2. Identify objectives.
3. Select training methods.
4. Implement the training program.
5. Evaluate, follow up the results of training.

As to training methods several are available (Schermerhorn, p 379). On-the-job training is done in the work setting while someone is doing the job. Some of the approaches include job rotation, formal and informal coaching, apprenticeship, and modeling. Job rotation allows people to spend time working in different jobs and thus expand the range of their job capabilities.

Coaching occurs when an experienced person gives specific technical advice to other workers; this can be done in a formal or a spontaneous setting . Formally when a supervisor explains otherwise when a coworker offers his expertise in case of need.

Apprenticeship involves a work assignment where someone serves as an understudy or assistant to a person who already has the required job skills. The apprentice learns a job over time, when he has a good example.

Modeling (role model) after a respected manager, supervisor enhances the skills of the understudy.

Off-the-job training is accomplished outside the work setting. It may be done within the organization at a separate training room or facility, or at an off-site location: the special training programs at universities, trade or professional associations, consultants. Management development is geared at enhancing the manageral skills (supra). These skills need continuous development. Mentoring where a senior manager takes responsibility for a junior manager, can be an important form of management development. Management simulation games give the opportunity to learn to handle complex problems. The same approach is taken in the virtual company where decisions provide insights in the actual company direction.

An often neglected source of training are business suppliers that offer technical courses and productivity and efficiency enhancing seminars. The impact on the bottom line is explained of the use of their products or machinery. Valuable insights in management techniques can be gained by scanning business offers, catalogues... Training on the premises of the material supplier or research in the laboratory of a cooperating partner gives the opportunity for benchmarking and/or acquiring basic skills.

Feasible options for Russian companies anno 1996/97 are: on-the-job training under the supervison of managers formed in the skills of management and market economy, these may come from the region (indigenous talent) or from abroad. Some Russian managers were exposed to the functioning of the market and acquired market skills since 1985 via studies and work experience abroad, TACIS programs...

Another method could be "training consultancy" or consultants, older managers and specialists from abroad that mentor the local talent. The advantage of on-the-job training is the direct applicability of the skills to the company.

Off-the-job training is possible at the business school as e.g. LETI-Lovanium that organize specialized programs. Training through international cooperation on the bilateral level (eg. the British Know-how Fund) and the multilateral level e.g. TACIS, EBRD, WCO... A massive transfer of management knowledge has taken place in the period 1990-1995 and anno 1996 several of these programs have formed some excellent managers and trainers that can compare the Russian practices with international standards in management education and organization learning.

All the universities are developing and some have already excellent undergraduate and graduate programs in business in place, however most of these programs – seen the financial situation of the government – are underfunded.

As a consequence companies must be willing to pay to get the necessary education for their managers.

To get an optimum return on training investment the following should be born in mind when it comes to deliver management education and how to approach the trainees in management.

FROM	TO
Teaching Process	Management Development Management Consulting and Action Learning
Institute-Based Teaching	Company-Based Learning
Trainer-Controlled Teaching	Learner-Controlled Learning, Self-Development
Standardized program	Tailor-made, Result-Oriented training Systems
Fragmented Training	Integrated into Company objectives Training

Source: Joseph Prokopenko, "Human Resources Management in Economics in Transition: The East European Case", Man Dev/66, ILO, 1992

Fig. 10. Training Approaches

The better training situations are in the right part of table 10, however in the near future any training is welcome.

It is obvious that the closer the instruction is to the company, the more tailor-made it is, the more expensive it will be, and the more dramatic the results ought to be = training consultancy. This is the ideal situation.

On the other hand, a lot of small, upstart enterprises cannot pay the high consultancy fees and have to use the open courses offered by training institutes, universities, incubation centers, trade associations, chamber of commerce...

7. Recommendations

The recommendations put forward are not all emcompassing but have the objective to serve as a base for reflection upon which action can be based.

7.1. The single most important production factor that determines your business success will be your workforce. Education, nurturing of this creative potential and the motivations of employees are of the utmost importance. Their knowledge, innovative potential, entrepreneurship and their ability (willingness) to give the customers value for money will be the critical succesfactor in a superautomated environment where perfect dissemination of information on a global basis is available. Computers and communication will eliminate barriers but also intensify competition. Doing better than the competition is the message and a prerequiste for company growth, survival and well-being. Investing in talent, people, their education and training is needed. Investment in training is an ongoing process since capabilities are getting obsolete very quickly.

7.2. Teambuilding and empowerment are two of today's important trends.

Managers create and improve systems that are operated by individuals and teams, mostly cross-functional teams. These are teams with members of the diverse functions (marketing, production...) on identical levels across the organization. The manager shares his power and authority with the teams so that they can make and implement decisions. Empowerment is found in flat organizations (with few hierarchical levels) and an important system in organizations that work with highly trained individuals and scholars and those that are on the cutting edge of technology.

A team should be cohesive and increase the efficiency of the company. Especially in science, innovation and for the creation of new products, the establishment of multifunctional teams is a prerequisite.

The art of teambuilding, working in teams, creating group cohesiveness, risk taking in groups and the group output has to be stimulated.

7.3. Total quality management as a company strategy should be applauded and will become a necessity when more buyers demand the ISO certification from their suppliers. Obtain the ISO-certification, overhaul the companies operations, make workers aware of quality and that it is only quality products that can increase the profit. TQM increases continuously the value to customers by designing and improving organizational processes and systems, it also involves all employees, suppliers and customers.

Training the employees in TQM and efficiency will increase the productivity and hence the bottom line.

7.4. Design competence profiles for all the functions in the company and recruit the people that match the needs of the company and fit the competency profiles.

7.5. Benchwork by visiting local companies that are front runners in their business or by contacting foreign companies. Ask a lot of questions on their organization, way of doing business, systems, competency profiles... and apply the new knowledge in the own company.

7.6. Form strategic alliances with local or foreign partners. In research and development this could avoid costly mistakes as to industry standards and could speed up the innovation and bring new products to the market in an expedient way.

7.7. Network and acquire new ideas and insights at conferences, training sessions, trade fairs, visits, workshops discussions with colleagues at home and abroad. This is an effective tool to disseminate knowledge on processes, systems, research...

8. Conclusions

The general educational level of the Russian population is very high indeed. There are also very intelligent and capable Russian Managers, Charismatic leaders and risk taking entrepreneurs (sometimes they take very high risks), but there are not enough of them yet. Massive training is requested in the near future, as are partnership – alliances in production, research, marketing, distributing, exports... that create a win-win situation for all parties involved.

Whereas the evolution in management thinking and philosophy in the Western world was slow and took half a century, progress in Russia and adaptation to the competitive environment is are equally enormous. The most modern technology is imported and up to-date management tools are used and excellent trainers are formed. But the effort should be sustained in the long run. The next decade will reveal whether the massive training and the change that is taking place will allow Russia to close the gap with its competitors and even enable it to surpass them and bring a vast selection of high quality techniques and products on the world markets. Integration in the world market by intra industry flows and investments will become apparent. Impediments come more from the political area and the public authorities than from the business world. When red tape can be cut, when privatization also means governance and control by the managers, when the banking sector gets stabilized the future looks promising indeed for business. Investing in education is the source of all progress for companies and gives a high return on investment especially in the high tech, science and R&D sector.

Bibliography

- Angehrn Albert, Doz Yves and Atherton Jill (1995). "Business Navigator – The next generation of management development t ools". EFMD Review, № 1, p. 24-31.

- Bentley David (1994). "Plugging West Central Gaps". EFMD Review, 1, p. 46-48.

- Brock Kay (1994). "Management Development in Central and Eastern Europe: Experience from Phare". EFMD Review, 94/3, p. 27-29.

- Chadwick John (1993). "Sundridge Park and the Russians-learning from each other". EFMD Quarterly Review, 2, p. 43-45.

- Claret Nathalie (1993). "L'intervention des consultants en entreprise. Direction et gestion des entreprises. № 140, mars-avril, p. 37-44.

- De Woot Philippe (1996). "The need for a sustainable learning society. EFMD Review, 2, p. 32-35.

- Derycke Luc (1996). Human Resource Management in Europe. IPO Management School - UFSIA, Working Paper, 17 p.

- Filatotchev Igor, Hoskisson Robert E., Buck Trevor, Wright Mike (1996). "Corporate Restructuring in Russian Privatization: Implications for US-investors". California Management Review, Vol. 38, № 2, Winter, p. 87-105.

- Giebens Guida (1996). Universitair Bedrijvencentrum Antwerpen: ontstaan ern eerste ervaringen. Economisch en Sociaal Tijdschrift, № 3, September, p. 501-527.

- Gottinger Hanss W. (1995). "Trends to support the ten commandments of general management: A European Perspective". EFMD Forum, 3, p. 29-31.

- Graham John L., Evenko Leonid I. (1992). "An Empirical Comparision of Soviet and American Business Negotiations". JIBS, Vol. 23, № 3, Third quarter, p. 387-419.

- Hooghiemstra Tjerk. "Management" of Talent". European Management Journal, Vol. 8, № 2.

- "Information Technology" (1995). EFMD Review, 1, p. 60.

- Katz Robert L. (1974). "Skills of an effective administrator". Harvard Business Review, vol. 52, September-October, p. 94.

- Keller A. Deborah and Campbell John F. (1992). Building Human Resource Capability. Human Resource Management, Vol. 31, Spring/Summer, № 1-2, p. 109-126.

- Kobychew A.N. (1995). Nordwestliche Region Ruslands Fihringskrafte des Unternehmertums und ihre qualifizuring bei der Wirtschaftlicher Transformation, St. Petersburg, 85 p.

- Kozminski Andrzej (1996). Succesfull Transition and Managerial Challenges. Annual Ceeman Conference, Praag, 12 p.

- Kraljic Peter Dr. (1996). Competitive global imperatives in Eastern, Central and Western Europe: Are they the same or different. Annual Ceeman Conference, Prague, McKinsey & Company, 60 p.

- Landa Ondrej (1995). "The new Europe. The Impact of Key Trends and discontinuities". EFMD Review, 2, p. 14-17.

- Lorange Peter (1996). "Developing learning Partnership". EFMD Review, 1, p. 29-33.

- Lundy Olive and Cowling Alan (1996). Strategic Human Resource Management. Routledge, London, 388 p.

- Manackovsky Andrei (1993). "Russian Management: How far from the West". EFMD Quarterly Review, 2, p. 28-33.

- "Management Development: Creating the future" (1995). EFMD Review, 2, p. 52.

- Management development in East-West Joint Ventures (1993). United Nations, New York, p. 128.

- Maruyama Magorah (1995). Misconceptions Regarding Communist Regime and Post Communist Reforme in Russia. Human systems Management, 14, p. 263-265.

- Mayo Andrew (1996). "The transfer of Management Wisdom across borders". EFMD Review, 1, p. 8-10.

- McCarthy Daniel J. and Puffer Sheila M. (1995). "Diamonds and Rust, on Russia's Road to Privatization: The profits and pitfalls for Western Managers". The Columbia Journal of World Business, Fall, p. 56-68.

- Mintzberg Henry (1994). "Rounding out the Managers' Job". Sloan Management Review, Fall, p. 11-26.

- Naughton Michael J. Bausch Thomas A. (1996). "The integrity of a Catholic Management education". California Management Review, Vol. 38, Summer, p. 118-140.

- Oakey Ray, Rothweel Ray, Cooper Sarah (1988). Management of Innovation in High Technology Small Firms, Pinter Publishers, 196 p.

- Poor Jozsef (1995). Human Resources Management. Practices under change in Central and Eastern Europe. CEEMAN, St. Petersburg, 20 p.

- Prokopenko Joseph (1992). "Human Resources Management in Economics in Transition: The East European Case", Management Development, 66, ILO 1992.

- Rada Juan (1992). "Quality in Management education". EFMD Review, 1, p. 25-26.

- Sattelberger Thomas (1992). "Towards the learning organization: the in-house business school of Deutsche Aerospace AG". EFMD Quarterly Review, 1, p. 5-8.

- Schermerhorn John R., Jr. Management for Productivity, Fourth Edition, John Wiley & Sons, USA, 758 p.

- Sharma Aviaham. Entry Strategies of US. Firms to the Newly Independent States, Baltic States and Eastern European Countries. California Management Review, Vol. 37, № 3, Spring, p. 90-106.
- Slatter Stuart (1992). Gambling on growth. How to manage the small High-Tech firm, John Wiley & Sons, Chichester, 319 p.
- Stahl, Michael J. (1995). Total quality Management in a global environment. Basil Blackwell Publishers, Oxford, 546 p.
- Twiss Brian (1987). Managing Technological Innovation, Pitman Publishing, 238 p.
- Van Weele Arjan J. (1996). Ondernemen en inkopen in Wild Water. IPO Management School, Paper, 14 p.
- Vecseny Janos and Baross Szabolcs (1996). Surviving Dinosaurs: Successful Firms in Transition. Working Paper, International Management Center, Budapest, 27 p.

Participants

Mr. Sergey A. BALANIOV
St.Petersburg Foundation for SME
Development, RUSSIA
Phone: (+7) 812-251-76-77,
Phone: (+7) 812-325-83-51
Fax: (+7) 812-112-66-07

Ms. Tatiana V. BAKHVALOVA
St.-Petersburg State Technical
University, RUSSIA
Phone: (+7) 812-552-95-32

Mrs. Ekaterina B. BESSARABOVA
The Analitical Centre of Science and
Industry Policy
Moscow, RUSSIA
Phone: (+7) 095-163-32-68
Fax: (+7) 095-163-32-68,

Mr Vadim G. CHYUMAK
International Institute of the Market,
Samara, RUSSIA
Phone:(+7) 846-35-18-13,
(+7) 846-35-72-39
Fax: (+7) 846-34-35-19

Mr. Cem ERGÜN
AREND Industrial Consultancy
TURKEY
Phone: (+90)-216-350-7341
Fax: (+90)-216-350-7341

Mr. Nikolay N. ERMILOV
Join Stock Enterprise ILIP
Phone:(+7) 812-234-37-95
Fax: (+7) 812-234-57-90

Dr. Gisella Meyer ESCOE
University of Cincinnati
Suncrest Dr, Cincinnati,
Ohio 45208, USA
Phone: (+1)-513-556-30-23
Fax: (+1)-513-556-29-53

Mr. Boris L. FEDOROV
Opened Business School
St.Petersburg Department,
RUSSIA
Phone: (+7) 812-311-27-48
Fax: (+7) 812-311-27-48

Prof. Andrei A. FURSENKO
St.Petersburg Regional Foundation
for Scientific and Technology
Development, RUSSIA
Phone:(+7) 812-394-35-83
Fax: (+7) 812-394-78-87

Mr. Manfred GARDYAN
Deutsche Management Akademie
Niedersachseng GmbH,
GERMANY
Phone: (+49)-51-41-97-32-10
Fax: (+49)-51-41-97-32-00

Mr. Igor V. GONNOV
"Obninsk-Technopolis" Association,
Obninsk, Kaluzhskaya obl., RUSSIA
Phone: (+7) 08-439-2-55-94
Fax: (+7) 08-439-4-82-25,
(+7) 095-230-23-26

Mr. Alexei A. GORDIENKO
Philosophy and Civil Law Institute
Siberian Department of Russian
Academy of Science (SD RAS)
Centre of Social Adaptation and Hihg
Quality Staf Training,
Novosibirsk, RUSSIA
Phone: (+7 383-2) 39-63-49,
(+7 383-2)-35-31-51

Mr. J.C. HENNEMAN
Consulate General of the Netherlands,
NETHERLANDS
Phone: (+7) 812-554-49-00
Fax: (+7) 812-554-36-19

Mr. Vladimir V. IVANOV
The Ministry of Science and
Technological Policy of
the Russian Federation
Moscow, RUSSIA
Phone: (+7) 095-229-26-80,
(+7) 095-229-17-00
Fax: (+7) 095-229-33-72

Mr. Hans-Joachim JOHN
AUCOTEAM - Ingeniergesellschaft
fur Automatisierung- und
Computertechnik mbH,
Berlin, GERMANY
Phone: (+49)-30-421-88-485
Fax: (+49)-30-423-27-09

Mr. Oleg S. KABANOV
St.Petersburg High-Tech Management
Training Center
St.-Petersburg, RUSSIA
Phone: (+7) 812- 394-35-83
Fax: (+7) 812- 394-79-51,
(+7) 812- 394-78-87

Mr. Alexey Yu. KERKIS
Deputy of R&D Administration
RD RAS
Novosibirsk, RUSSIA
Phone: (+7 383-2)-35-32-40
Fax: (+7 383-2)-35-77-69

Mr. Peter A. KISELEV
German-Russian Centre for technical
Assistance Transform
ST.-PETERSBURG, RUSSIA
Phone: (+7) 812-27-89-09
Fax: (+7) 812-27-89-11

Mr. Alexander N. KOBYSHEV
"Progress" Management and
Marketing Centre
ST.-PETERSBURG, RUSSIA
Phone: (+7) 812-552-13-38
Fax: (+7) 812- 552-77-28

Mrs. Sveta KOCHARNAIA
Trans National Projects (TNP)
Management
BELGIUM
Phone: (+32) 33-24-51-65
Fax: (+32) 33-24-51-65

Mr. Ustin V. KOLBIN
Sciense–Technological Park of
St.Petersburg
St.-Petersburg, RUSSIA
Phone: (+7) 812- 172-38-51
Fax: (+7) 812- 172-38-51

Mr. Tadeusz KOZEK
Task Force for Training and Human Resources Cooperation Fund
POLAND
Phone:(+48)-22-625-28-61,
(+48)-22-625-39-37
Fax: (+48)-22-625-28-05

Mr. Konstantin V. KOZLOV
PLC AVANGARD
St.-Petersburg, RUSSIA
Phone: (+7) 812- 543-7678,
(+7) 812- 540-9155
Fax: (+7) 812- 540-1844

Dr. Jochem KREHER
BBZ GmbH
GERMANY
Phone: (+49)-30-69-52-41-95,
Phone: (+49)-30-69-52-40-15,
Fax: (+49)-30-69-52-40-14

Mr. Finn KROGH
Carl Bro International a/s, TACIS
DENMARK
Phone: (+7) 812- 325-8351,
(+7) 812- 251-7677
Fax: (+7) 812- 112-66-07

Mr. Lev V. KUZTETZOV
Sciense & Coordinative Center "PERSPECTIVA"
Samara, RUSSIA
Phone: (+7) 846 242-31-01
Fax: (+7) 846 232-11-53

Ms. Ekaterina M. KUZNETZOVA
St.-Petersburg State Technical University
St.-Petersburg, RUSSIA,
Phone: (+7) 812- 552-95-32

Mr. Herman Louis LEIJDESDORFF
CAP VOLMAC, HOLLAND
Phone: (+31)-30-692-98-40
Fax: (+31)-30-691-43-82
Fax: (+31)-30-691-48-87

Mrs. Lidia S. LEONTIEVA
Regional Centre for Training of Personnel
Yaroslavl, RUSSIA
Phone: (+7) 0852-32-88-36
Fax: (+7) 0852-32-83-24,
Fax: (+7) 0852-32-83-27

Mr Valentin N. LEVITSKIY
The Russian Technology Fund L.P.
St.-Petersburg, RUSSIA
Phone: (+7) 812-311-77-58
Fax: (+7) 812-311-77-58

Mr. Mikhail A. MAIOROV
St.-Petersburg State Technical University
St.-Petersburg, RUSSIA
Phone: (+7) 812- 552-95-32

Mr. Luigi MASSIMO
Service R&D Cooperation with
Central and Eastern Europe
Commission of the EC
Directorat-General – SR&D,
BELGIUM
Phone: (+32)-2-295-66-49
Fax: (+32)-2-296-33-08

Mr. Vladimir N. MIKHAYLOV
Novgorod Sciense–Coordination
Center
Novgorod, RUSSIA
Phone: (+7) 816-7-81-25

Mr. Fritz MUELLER
Siemens Nixdorf, MUENCHEN
GERMANY
Phone: (+49)-89-63-64-42-60
Fax: (+49)-89-63-64-31-57

Mrs. Dorina MUNTEAN
Trans National Projects (TNP)
Management
Bucharest, ROMANIA
Phone: (+40) -1-33-603-44,
(+40)-1-33-630-69
Fax: (+40)-1-33-302-66

Mr. Hanns Peter MUTH
Enterprise PLUS Ltd.,
GREAT BRITAIN
Phone: (+44)-4-42-87-03-91
Fax: (+44)-442-87-03-45

Mrs. Albina I. NIKKONEN
St.Petersburg High-Tech Management
Training Center, St.Petersburg
RUSSIA, 197348, Box 26
Phone/Fax: (+7) 812-394-78-87
Fax: (+7) 812-394-79-51

Mrs. Nina Y. ODING
"THE LEONTIEF CENTRE"
16 Voznesensky Ave. St. Petersburg,
190000, RUSSIA
Phone: (+7) 812-319-41-19,
(+7) 812-319-91-00
Fax: (+7) 812-319-98-14

Mr. H.H. von OERTZEN
BBZ GmbH, GERMANY
Phone: (+49)-30-69-52-41-95,
Phone: (+49)-30-69-52-40-15,
Fax: (+49)-30-69-52-40-14

Mr. Vicharn PANICH
The Thailand Research Fund
THAILAND
Phone: (+622)-64-25-18-69
Fax: (+662)-64-25-1-90

Mr. Gennady F. PECHENITSIN
"Perspectiva" Professional
Training-Centre
St.-Petersburg, RUSSIA,
Phone: (+7) 812-144-04-64
Fax: (+7) 812-144-79-84

Mr. Konstantin PLETNEV
Regional Centre of R&D Co-operation, Presidium of RAS,
"RENATECHS"
Moscow, RUSSIA
Phone: (+7) 095-492-83-61

Mr. Gennadiy S. POSHEVNEV
Novosibirsk a Urban Centre of
Employment of the Population
NOVOSIBIRSK, RUSSIA
Fax: (+7) 383-222-00-26

Mrs. Sarah POWELL
Know How Fund, British Consulate-General, St.Petersburg
GREAT BRITTAN
Phone: (+7) 812-325-60-36
Fax: (+7) 812-325-60-37

Mr. Dmitriy V. SERGEEV
First Deputy Governor of
St.Petersburg
Phone: (+7) 812-315-51-52
Phone: (+7) 812- 310-05-05
Fax: (+7) 812- 113-48-25

Mrs. Yelena G. SLOMTCHINSKAJA
Laser Association,
Moscow, 117485
Box 27, RUSSIA
Phone: (+7) 095-923-53-75;
Fax: (+7) 095- 924-87-42

Mr. Alexander V. SOKOLOV
Sciense Research & Statisrics Center
Moscow, RUSSIA
Phone: (+7) 095-229-82-27
Fax: (+7) 095-924-28-28

Mr. Vladimir I. SPIVAK
St.Petersburg Regional Foundation for
Scientific and Technology
Development, RUSSIA
Phone/Fax:(+7) 812- 394-78-87
Phone: (+7) 812- 394-35-83

Mrs. Irina A. SUVOROVA
Research & Consulting firm ALT
St.-Petersburg, RUSSIA
Phone: (+7) 812- 213-99-81
Fax: (+7) 812- 213-32-59,
Fax: (+7) 812- 217-54-79,

Mr. Hans-Jochem SMIDT
Consulate General of German
Federative Respublic
GERMANY
Phone: (+7) 812-273-55-98,
(+7) 812-273-57-31
Fax: (+7) 812-279-32-42

Prof. Alexander P. SOROKIN
"Obninsk-Technopolis" Association
Obninsk, Kaluzhskaya obl.,
RUSSIA
Fax: (+7) 08-439-4-82-25,
(+7) 095-230-23-26

Mr. Viacheslav V. VAKHOVSKY
Ulyanovsk Business-Centre BINK
Goncharnaya str., 36
Ulyanovsk, RUSSIA
Phone: (+7) 8422-31-93-26
Fax: (+7) 8422 63-69-69

Prof. Liliane VAN HOOF
University of Antwerpen (UFSIA)
NETHERLANDS
Phone: (+32)-3-220-41-21
Fax: (+32)-3-220-47-99

Mr. Emilio VENTO
UNIDO United Nation Industrial
Development Organisation
AUSTRIA
Phone: (+43)-1-211-31-37-26
Fax: (+43)-1-211-31-68-11

Mr. Carolus M.R. VERKOEYEN
Trans National Projects (TNP)
Management
NETHERLANDS
Phone: (+31)-181-41-89-60
Fax: (+31)-181-41-89-61

Prof. Alexander D. VICTOROV
Vice rector of St.Petersburg
Electrotechnical University
197376, Prof.Popov st., 5
St.Petersburg, RUSSIA
Phone: (+7) 812-234-15-43
Fax: (+7) 812-234-15-43

Mr. Wilfred Herbert WHITLOW
NIMTECH,
UNITED KINGDOM
hometel: (+44)-19-28-72-45-64
Phone: (+44)-17-44-45-33-66
Fax: (+44)-17-44-45-33-77

Mr. Alexander E. YANCHEVSKIY
LETI-Lovanium, Intrnational School
of Management, RUSSIA
Phone: (+7) 812-234-13-91
Fax: (+7) 812-234-02-54

Mr. Igor L. ZAKHAROV
Sciense & Education Departament
Administration
Nizhniy Novgorod, RUSSIA
Fax: (+7) 8312-33-37-24

Mr. Igor V. ZAITZEV
St.-Petersburg State Technical
University, RUSSIA
Phone: (+7) 812-552-95-32

Mr. Oleg E. ZHUKOV
St.-Petersburg State Technical
University, RUSSIA
Phone: (+7) 812-552-95-32

Index

MIX
Papier aus verantwortungsvollen Quellen
Paper from responsible sources
FSC® C105338

If you have any concerns about our products,
you can contact us on
ProductSafety@springernature.com

In case Publisher is established outside the EU,
the EU authorized representative is:
Springer Nature Customer Service Center GmbH
Europaplatz 3, 69115 Heidelberg, Germany

Printed by Libri Plureos GmbH
in Hamburg, Germany